W9-CKC-802

# 49 Electronic 6-Volt Projects

*Delton T. Horn*

*Trademarks*

Mylar is a trademark of E.I. Dupont de Nemours
Radio Shack is a registered trademark of the Tandy Corporation

FIRST EDITION
FIRST PRINTING



Library of Congress Cataloging-in-Publication Data

Horn, Delton T.
49 electronic 6-volt projects/ by Delton T. Horn.
p. cm.
Reprint.
ISBN 0-8306-8975-3 ISBN 0-8306-3275-1 (pbk.)
1. Electronics—Amateurs' manuals. I. Title. II. Title: Forty-nine electronic six-volt projects.
TK9965.H636 1990
621.381'078—dc20 89-27494
CIP

TAB BOOKS Inc. offers software for sale. For information and a catalog, please contact TAB Software Department, Blue Ridge Summit, PA 17294-0850.

Questions regarding the content of this book should be addressed to:

Reader Inquiry Branch
TAB BOOKS Inc.
Blue Ridge Summit, PA 17294-0214

Acquisitions Editor: Roland S. Phelps
Technical Editor: B. J. Peterson
Production: Katherine Brown

# Contents

# List of Projects

# Introduction

THIS BOOK FEATURES FORTY-NINE PROJECTS FOR THE ELECTRONics experimenter to build and enjoy. Some have serious applications, and others are just for fun. An unusual feature of this book is that the projects are all designed to operate from +6Vdc (volts direct current). This can be very convenient for you. You do not have to build a new power supply for each new project. You do not have to fool around with several different battery sizes. If you use a variable power supply, there is less chance that you will accidentally hook up a circuit with the power supply set for the wrong voltage. A 6V lantern battery is a good power supply for these projects.

Many of the projects in this book are ideal for science fair projects. Some prior experience with electronics project building is assumed, but most of the projects are simple enough for the beginner who knows which end of a soldering iron to hold.

In selecting the projects to include in this book, I tried to offer as much variety as possible. The 49 projects include *LED* (light-emitting diode) flashers, digital circuits, tone generators, timers, and many other useful and interesting electronic devices. You are encouraged to experiment with the circuits shown here and to modify them to suit your applications.

The choice of projects was made using three criteria:

- Is the project fun?
- Is the project useful?
- Can the hobbyist learn from this project?

I had fun creating the projects for this book. I hope you enjoy building them and experimenting with them.

If you enjoy this book, you will also enjoy the companion volume, *49 Electronic 12-Volt Projects* (TAB Books #3265).

# 1 Power Sources

THE BULK OF THIS BOOK IS MADE UP OF OVER FOUR DOZEN electronic projects. Each project is designed to operate from +6 Vdc (volts direct current). Many other project books can be frustrating when projects call for unusual and sometimes difficult to obtain supply voltages. Six volts is a very common value. Many batteries and commercial power supplies can be used to provide +6V to these projects. This chapter discusses some of the ways +6V can be supplied to your projects.

## BATTERIES

The most obvious and simplest source of dc power is a battery pack. Most standard batteries (AAA, AA, C, and D) put out 1.5V. To make up a 6V battery, you will need to use four 1.5V cells in series.

Lantern batteries, which supply 6V are fairly common. They put out a pretty hefty amount of current, so they are suitable for high-drain circuits or applications that require continuous power over long periods. The chief disadvantage of lantern batteries is that they are large. They are heavy and bulky, which could present a problem in applications that must be portable.

## AUTOMOBILE POWER

Most automobiles use 12V batteries. This voltage can be tapped off at various points and used to power electronic circuits. To obtain 6V from an automotive 12V battery, you will need to use a voltage regulator or a resistive voltage divider. A few automobiles use a 6V system, but these systems are becoming increasingly rare.

The simplest approach is to take the 12Vdc from the cigarette lighter. Special adapters for this purpose are available from Radio Shack and many other dealers.

## AC-TO-DC CONVERTERS

If you will be using a project exclusively in an area where ac (alternating current) house current is available, it makes sense to tap this power source. It is certainly cheaper than batteries. However, these projects are designed to run off 6Vdc, and house current is nominally 110Vac. Obviously, you need a conversion circuit.

## PROJECT 1: 6-VOLT POWER SUPPLY

A simple "quick-and-dirty" 6V power supply circuit appears in Fig. 1-1. A parts list for this circuit is given in Table 1-1. This power supply can be used to run a noncritical circuit.

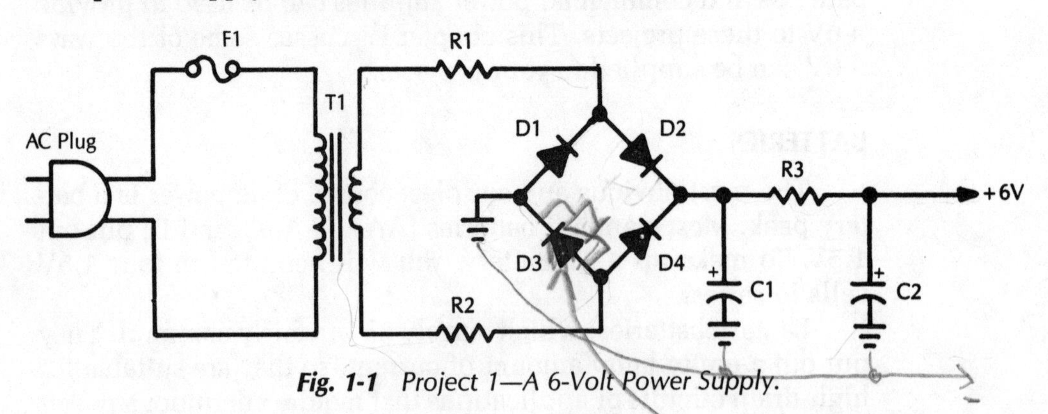


*Fig. 1-1* Project 1—A 6-Volt Power Supply.

**Table 1-1 Parts List for the 6-Volt Power Supply.**

| Part | Component Needed |
|---|---|
| D1 – D4 | 1N4002 diode |
| T1 | Power transformer, secondary 8 – 10V (approximate) |
| F1 | Fuse to suit load |
| C1, C2 | 500$\mu$F (micro-farad) electrolytic capacitor |
| R1, R2, R3 | 1k$\Omega$ (ohm) 2W (watt) resistor |

## PROJECT 2: REGULATED 6-VOLT POWER SUPPLY

Some circuits, especially those involving digital circuits, require a regulated dc supply voltage. A regulated 6V power supply circuit appears in Fig. 1-2, with the parts list given in Table 1-2.

## SAFETY

With any ac-powered circuit, always use a fuse or circuit breaker. These safety devices interrupt the ac input to the circuit if the current drawn becomes too great. If you do not use a fuse or circuit breaker, you could get a painful and dangerous (and possibly even fatal) electrical shock.

Select the fuse or circuit breaker rating for the circuit to be powered. Estimate the maximum amount of current the device should normally draw and add about 10 percent as a safety factor. For example, if a circuit will normally draw up to 1.1A (amperes) adding 10 percent results in 1.21A. In this case use a 1.5A fuse, which is the nearest standard-size fuse.

In an existing device, never replace the existing fuse or circuit breaker with one having a higher rating. If the fuse blows repeatedly (or if the circuit breaker trips repeatedly), something is wrong. Disconnect power and troubleshoot the problem. Do not bypass the fuse or circuit breaker. Do not use too large a fuse or circuit breaker. If you use too large a current rating for the fuse or circuit breaker, some component in the circuit—usually a relatively expensive transistor or IC (integrated circuit)—will burn out to protect the fuse or circuit breaker. Obviously, there is not much point to that. Do not defeat the purpose of the safety device.

Make sure that all conductors carrying ac house current are well insulated. Be sure it is not possible for anyone to accidentally touch a live conductor.

Also follow any special safety precautions described for individual projects.

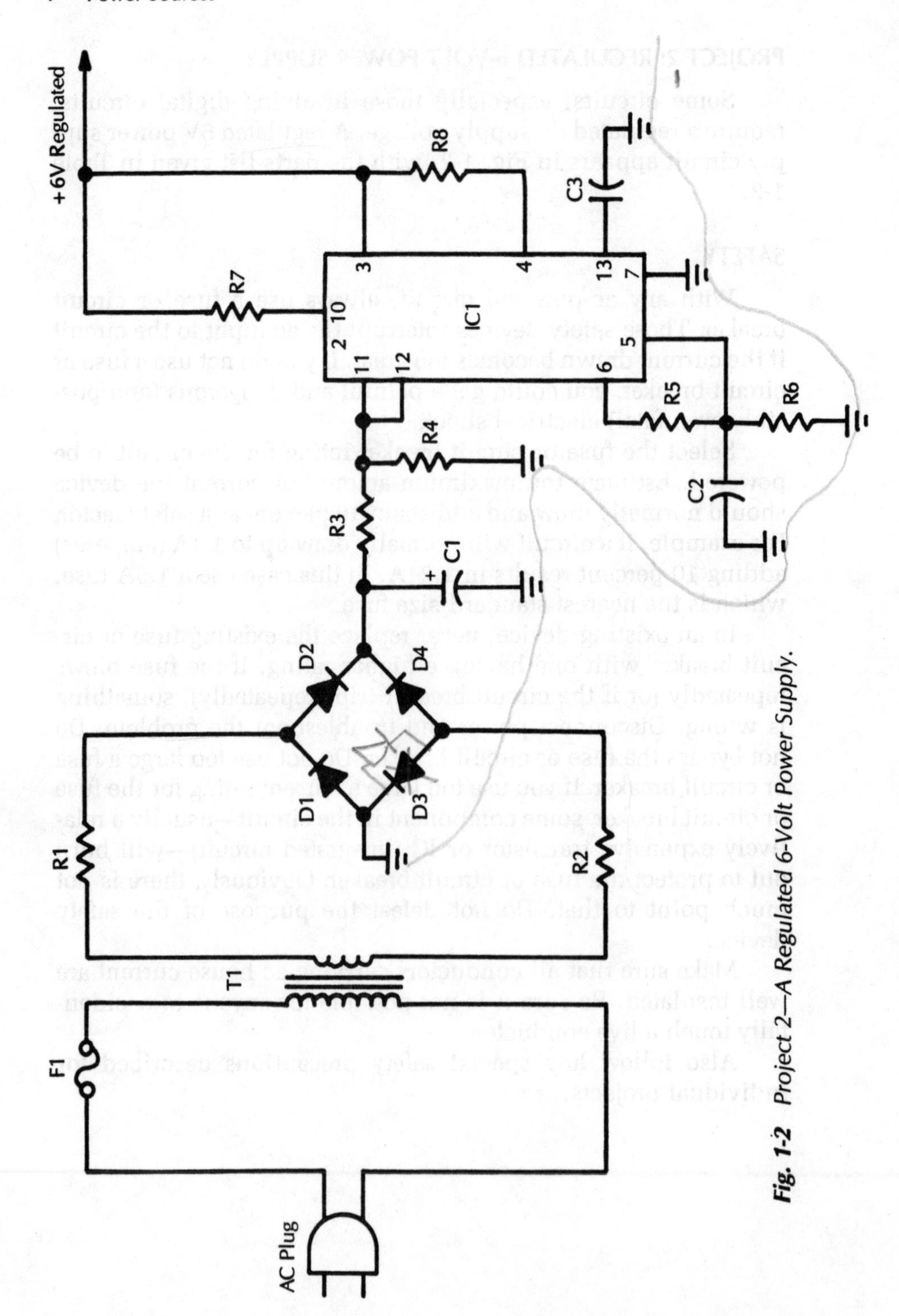


**Fig. 1-2** *Project 2—A Regulated 6-Volt Power Supply.*

**Table 1-2 Parts List for the Regulated 6-Volt Power Supply.**

| Part | Component Needed |
|---|---|
| IC1 | 723 adjustable voltage regulator |
| D1 – D4 | 1N4002 diode |
| T1 | Power transformer, secondary 10 – 12V (approximate) |
| F1 | Fuse to suit load |
| C1 | 500μF electrolytic capacitor |
| C2 | 0.01μF capacitor |
| C3 | 100pF (pico-farad) capacitor |
| R1, R2 | 1kΩ 2W resistor |
| R3 | 10kΩ 1W resistor |
| R4 | 2.2kΩ 1W resistor |
| R5, R8 | 1kΩ 1/2W resistor |
| R6 | 6.2kΩ 1/2W resistor |
| R7 | 22Ω 1/2W resistor |

# 2 ❖

# Construction Tips

THIS CHAPTER IS A BASIC INTRODUCTION TO ELECTRONICS PROJECT building for the beginner. If you are an experienced hobbyist, you might also want to read this introductory chapter as a convenient review. In addition to the basics of project constructions, this chapter also covers the various power supply options you can use to run your projects.

## BREADBOARDING

You could just build the projects directly, constructing them on simple perforated boards with point-to-point wiring or on simple homemade printed-circuit boards. However, before you reach for your soldering iron, I strongly suggest that you *breadboard* (temporarily connect the components without solder) each project first. This approach is a good idea for almost any electronics project. It is much easier to troubleshoot a problem or make modifications to a circuit if you do not have to waste a lot of time desoldering and resoldering connection points. As well as being time consuming and a real nuisance, desoldering and resoldering can introduce a number of new problems. *Solder bridges* (creating short circuits between adjacent leads or connection points) become increasingly likely every time you apply the soldering iron. Reheated solder (during resoldering) is especially prone to flowing someplace it should not go. In addition, excessive heat can all too easily damage or destroy some components—especially semiconductors, which are inherently sensitive to heat and are somewhat delicate. It is obviously best to apply heat as infrequently (and for as short a time) as possible.

Breadboard the circuits first. When you are 100 percent sure that everything is right and that the circuit does exactly what you want it to do, then solder a permanent version of the project. Breadboarding might seem like an unnecessary extra step, but in the long run, it can really save you a lot of time, frustration, and even money.

Using a solderless socket, like the one shown in Fig. 2-1, to breadboard a circuit also encourages experimentation, which is unquestionably the best way to learn about electronics. In some cases, you might not need or want a permanent version of certain projects. You might only want to experiment with a circuit for a while. In such cases, a solderless breadboard is the only reasonable way to go. There will be a minimum of bother, and you can easily reuse the components over and over.

You can use a solderless socket by itself to breadboard your circuits, but this method leaves something to be desired. For one thing, a lot of extra work can be involved. Many projects require external circuits. Typical examples of external circuits are power supplies (which are discussed in this chapter) and oscillators (to serve as signal sources). It is a nuisance to have to build such a basic, common circuit from scratch every time you want to breadboard a project.

A second disadvantage of the solderless socket is that it is awkward at best to work with any large components (such as

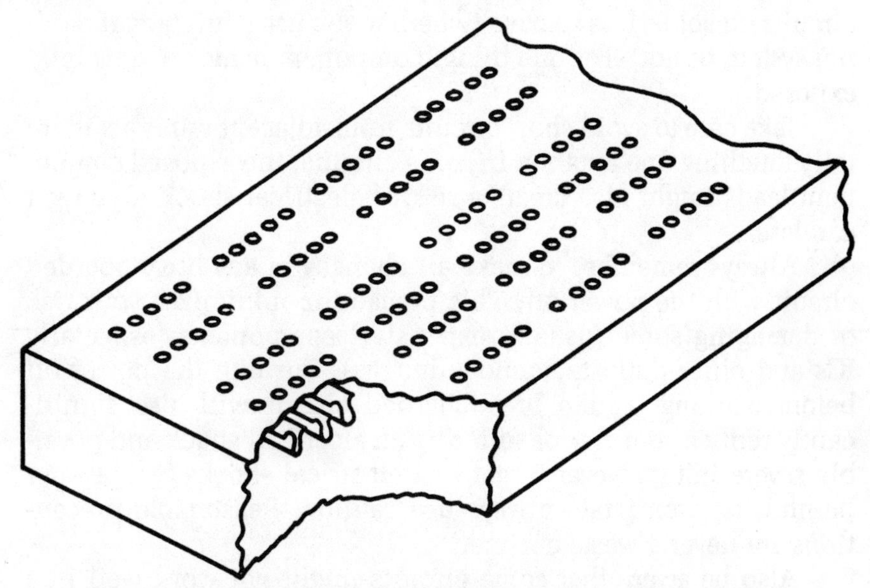

***Fig. 2-1*** *A solderless socket simplifies circuit experimentation.*

switches, potentiometers, transformers, and the like). These large components cannot be mounted in the small holes of the solderless socket.

The solution to both these problems is to use a *full breadboarding system*. Such a system is built around a solderless socket, but it is more than just the socket itself.

Mount one (or sometimes more than one) solderless socket on a secure base. The base is actually a housing for several common circuits (power supplies, oscillators, amplifiers, etc.), which may be frequently needed when breadboarding projects. Convenient connection points are provided to run wires from these circuits to the breadboarded circuit being built on the solderless socket.

Also mount several common large components in the base. These components might include one or more switches, one or more potentiometers, a speaker, and possibly a transformer, LED or *LCD* (liquid-crystal display) unit.

You can buy ready-made breadboarding systems from a number of manufacturers, or you can design and build a customized breadboarding system of your own. Coming up with your own system is an especially good idea if you do a lot of work with a particular type of project and have special requirements.

There are some precautions involved when breadboarding circuits on solderless sockets (whether you use a full breadboarding system or not). For one thing, component leads are generally exposed.

Take care to avoid short circuits from adjacent wires accidentally touching one another. In some circuits, the exposed component leads might also create a risk of electrical shock if you get careless.

Always remember to make all changes in any breadboarded circuit with the power off. This precaution minimizes your risk of damaging some (usually expensive) components (especially ICs and other delicate semiconductors). Turning the power off before working on the breadboarded circuit will also significantly reduce your risk of suffering an electrical shock and possibly severe injury. Never forget that electrical shocks can be very painful, or even fatal—always use caution. Reasonable precautions are never a waste of time.

Also be aware that some circuits might not work well in a standard solderless breadboarding socket. In some rare cases they might not work at all. Such problems are most often encountered with circuits utilizing high-frequency signals. In high-frequency circuits, the length of connecting wires and shielding (or lack thereof) can be critical. In a breadboarded circuit, there can also be a problem with *phantom components* (stray capacitances and inductances). Such phantom components generally have very small values and thus have a negligible effect in most circuits. However, in high-frequency circuits, they might alter the way the circuit functions considerably.

Fortunately, the circuit that cannot be prototyped in a breadboarding socket, at least to a reasonable degree, is very much the exception to the rule. If high frequencies do happen to be involved in a project you are working on, you can minimize potential problems by keeping interconnecting wires as short as possible. If the breadboarded circuit operates incorrectly or erratically, try relocating some of the components. This will often (though not always) clear up the problem. You should not run into such problems with any of the projects in this book.

Some breadboarded circuits might change some of their operating parameters noticeably when the project is converted to a more permanent form of construction. The changes might be slight, or they might be considerable. Be aware of such potential problems when experimenting with a breadboarded circuit. There really is not very much you can do about this problem in advance, but at least be prepared to recognize such problems when they do crop up. If nothing else, you might be able to save some time and frustrated hair pulling.

Once you have breadboarded and experimented with your project and have gotten all of the bugs out, you will probably want to rebuild some circuits in a more permanent way. Solderless breadboarding sockets are great for testing and experimenting with prototype circuits, but they really are not much good when it comes to putting the project to practical use.

Breadboarded circuits, by definition, use temporary connections. In actual use, some component leads can easily bend and touch one another, causing potentially harmful short circuits. Components might even fall out of the sockets when the device is moved about. Interference signals can be generated easily and picked up by the exposed wiring.

Packaging a circuit built on a solderless socket will be tricky at best. They tend not to fit very well in standard circuit housings and boxes. Also, a solderless socket is fairly expensive. It is certainly more than worth the price if it is re-used for many different circuits. But if you tie it up with a single permanent project, you are only cheating yourself. Less expensive construction methods that are more reliable, more compact, and that offer better overall performance are readily available. Some of the more common permanent construction methods are discussed briefly in the next pages.

## PERFORATED BOARD

You can construct many relatively simple circuits, including most of the projects presented in this book, on a perforated circuit board, or *perf board*. A typical perf board is illustrated in Fig. 2-2. Mount component leads through the perforated holes in the nonconductive board. The holes are evenly spaced in rows, and most standard components can fit onto such a board. Solder component leads and jumper wires together directly, using point-to-point wiring.

*Flea clips* are used by many hobbyists. A flea clip is inserted through one of the holes in the perf board, and component leads are attached to the flea clip, rather than directly through the hole itself. See Fig. 2-3.

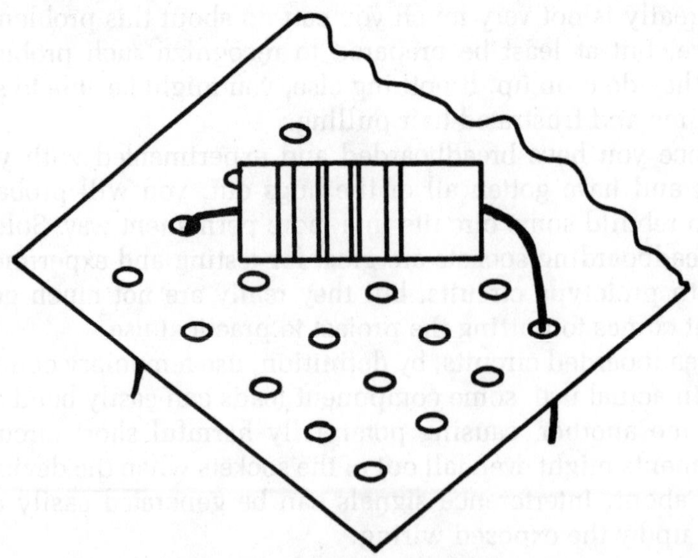

***Fig. 2-2*** *Component leads are mounted through the holes of a perforated board.*

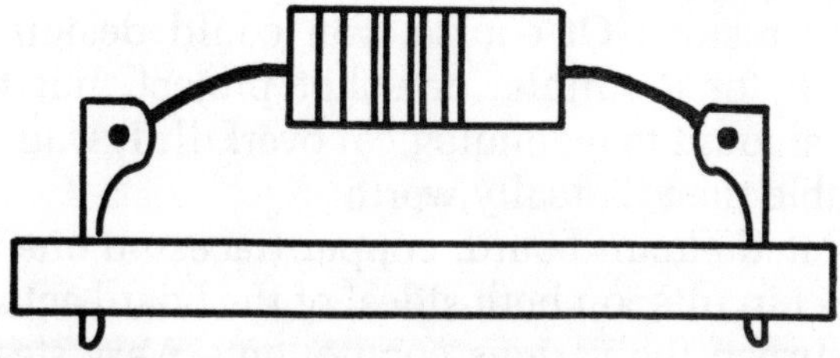

*Fig. 2-3 Flea clips are used by many hobbyists to hold component leads.*

The perf board serves as a secure base for the circuitry mounted on it. Wire only very, very small, very simple circuits directly together without any supporting circuit board. The board provides physical support and helps minimize problems of a *rat's nest* or jumbled wiring. A rat's nest is next to impossible to trace if an error is made in construction or if the project requires servicing at some later date. Rat's nest wiring is also an open invitation to short circuits and/or breaks in connecting wires. Loose hanging wires can create electrical problems, such as stray capacitances and inductances between nearby wires. Stray capacitances or inductances can allow signals to get into the wrong portions of the circuit. This will obviously cause erratic operation, if not complete circuit failure. In a few cases, such stray signals getting in the wrong place could conceivably cause permanent damage to certain components.

Although use of a perf board will automatically minimize some aspects of rat's nest wiring, it is still possible to use sloppy construction, resulting in a rat's nest of sorts. There are some simple tricks to minimize such problems. Arrange the components on the board before you begin soldering. Make sure everything fits neatly. Keep all connecting leads and jumper wires as short as possible. Avoid the use of crossed jumper wires wherever possible. In circuits of any complexity you will find it nearly impossible to eliminate all such wire crossings, but try to arrange the components so that as few wire crossings as possible are necessary. Of course, you must insulate adequately any wires that cross one another to prevent short circuits. Use straight-line paths for jumper wires wherever possible.

## PRINTED CIRCUITS

For moderate to complex circuits, or for circuits from which a number of duplicates will be built, use a printed circuit board for very good results. Of course, you could design and use a printed circuit for a simple, one-shot project, but that would pretty much amount to technological overkill. It would probably be more trouble than its really worth.

On a *printed circuit board,* copper traces on one side (or, in very complex circuits, on both sides) of the board act as connecting wires between the various components. Very steady, stable, and sturdy connections can be made because the component leads are soldered directly to the supporting board.

The original board is a slab of insulating material (like that used in perf boards), with one side (or sometimes both sides) covered with a thin layer of copper. The desired pattern of connecting traces is applied to copper-clad side(s) of the board.

You can draw the pattern on directly with special resist ink or use stick-on resist labels. More advanced experimenters and professional manufacturers usually apply the trace pattern with photographic techniques.

Once the pattern of desired copper traces is applied (by whatever means), soak the board in a bath of a special acid or etchant. The acid will eat away the exposed copper, but the resist ink or labels protect certain portions of the copper cladding. When you remove the board from the etchant bath and wash it to remove the excess acid and the resist ink, the only remaining copper will be in the form of the desired pattern of traces. Finally, drill holes to accommodate the leads of the components to be mounted on the board.

Take great care in laying out a PC (printed circuit) board. The exact sizes and positions of all components must be precisely worked out before the resist is applied to the board. If you make a mistake, you will probably have to do the entire board over. Take special care to avoid wire crossings. The copper traces cannot cross over each other (unless you are using a dual-sided board) because the traces can only exist in two dimensions. If a crossing is absolutely essential, use a wire jumper to connect two separated traces. Wire jumpers are mounted onto PC boards just like ordinary components.

Stray capacitances between traces can adversely affect circuit performance in some projects (especially when high-frequency signals are involved). In critical circuits, a guard band between traces can help reduce the potential problem of stray

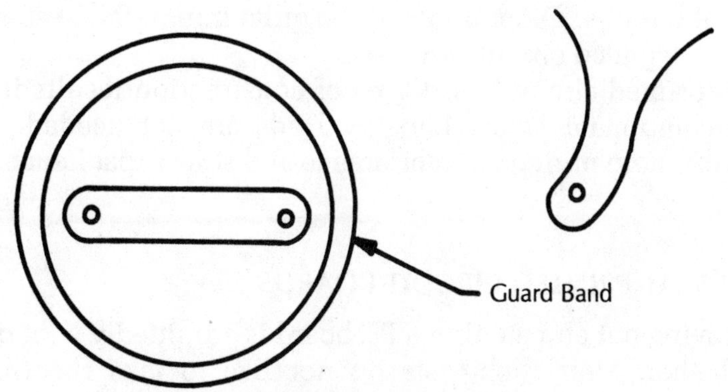


***Fig. 2-4*** *Guard bands can be included on a printed circuit board to minimize stray capacitances.*

capacitances. An example of the use of guard bands is illustrated in Fig. 2-4.

The copper traces on most PC boards are usually placed very close to one another. This is especially true in circuits using ICs, because the IC leads do not have much spacing between them. Because adjacent traces are close to each other, a short circuit is very easy to create on a PC board. A small speck of solder or a piece of excess lead from a component could very easily bridge across two (or more) adjacent traces, creating a short. When soldering component leads to a PC board, use a minimum of solder. If you use too much solder, it will tend to flow and bridge across adjacent traces.

Be careful not to apply too much heat when soldering on a printed circuit board. Do not leave the soldering iron at any one point for too long a time. Excessive heat will damage the bond between the copper cladding and the supporting board. The copper trace will tend to peel up and away from the board. The board will then be extremely fragile, and breakage will be almost inevitable.

Tiny, nearly invisible cracks in the copper traces can also be a problem, if you are not careful at all stages in PC construction. There are many ways such cracks can be caused. If you use reasonable care, you should not have many problems of this type, but occasionally they will crop up, even for the best of us. Generally, fairly wide traces that are widely spaced are the easiest to work with. Unfortunately, wide traces are not always practical with all circuits. Often they will be quite impossible, especially where integrated circuits are used.

A printed circuit board type of construction results in very short component leads. Lengthy leads are not needed. Short leads can help minimize interference and stray capacitance problems.

## UNIVERSAL PRINTED CIRCUIT BOARDS

Laying out and etching a PC board is admittedly a lot of fuss and bother. Many hobbyists do not like to take the trouble. Recently, a new choice has been added to the list of project construction methods. This is the use of a universal printed circuit board.

A *universal PC board* is a commercially available board with a standardized pattern pre-etched onto it. The pattern on the universal PC board can be used to build a great many different projects. A universal PC board is sort of a cross between a perf board and a dedicated PC board. Like a perf board, it has a number of rows of holes. Place components where convenient (as long as the leads go through to the desired trace). Most of the holes in the board are left unused. An ordinary PC board usually has no excess holes. Holes are specifically drilled in precise locations for the specific components used in the circuit.

Several universal PC board designs are available, including designs for analog, digital, and *op-amp* (operational amplifier)—analog with two power supply buses—circuits. Select whichever pattern is most convenient for your individual project. Radio Shack stores, and a number of other dealers, carry several universal PC boards.

## WIRE WRAPPING

Many circuits using several ICs are being constructed today using a method known as *wire wrapping*. A thin wire (typically 30 gauge) is wrapped tightly around a special square post. The squared off edges of the post bite into the wire, making a good electrical and mechanical connection without soldering. Components are fitted into special sockets that connect their leads to the square wrapping posts. Wire wrapping sockets are available to accommodate most IC *DIP* (dual inline package) packages.

The wire wrapping construction technique is most appropriate for circuits that are made up primarily of several integrated circuits. If just a few discrete components are included in the circuit, they can be fitted into special sockets, or headers. Alternatively, the discrete components can be soldered directly, while the connections to the ICs are wire wrapped. This is known as *hybrid construction*. In circuits involving many discrete components, the wire wrapping method tends to be rather impractical. Use a different method of circuit construction.

Wire wrapped connections can be made (or unmade) quickly and easily, without any risk of potential heat damage to delicate semiconductor components (ICs). This is because no heat source (soldering iron) is applied to the component leads.

You will certainly find some disadvantages to using this method to construct your projects. As mentioned above, discrete components (resistors, capacitors, transistors, etc.) are awkward at best. Also, the thin wire wrapping wire is very fragile and quite easily broken. Because the wire is so thin, it can only be used to carry very low power signals. Stray RF (radio frequency) pick-up and emission can be a problem, especially with high-frequency circuits. The wiring in a wire-wrapped circuit can be very difficult to trace. When a large number of ICs are involved (some circuits require several dozen), wire wrapping can be a very convenient way to construct a project.

Wire wrapping is described here merely for the sake of completeness. It would probably not be a particularly good choice for the construction of any of the projects presented in this book.

## SUBSTITUTING COMPONENTS

With very few exceptions (which will always be indicated in the text) the components used in these projects are not overly critical. Most projects need only readily available components. If you cannot get the exact component called for, you should be able to find a reasonable substitute without much trouble. If you already have something in your junk box that will do the job, there is certainly no point in buying a brand new part with a slightly different identification number.

Use a good substitution guide to find equivalent devices for any semiconductors, especially transistors. ICs tend to be more difficult to substitute. Electronics is a constantly changing field.

It is entirely possible that a chip that was widely available when this book was written might be discontinued by the time you read it. Your best bet in such cases is to try the various surplus houses. Many advertise in the backs of the electronics hobbyist magazines. Also check your local yellow pages. There might be a surplus house or a well-stocked jobber supplier near you.

All digital ICs are of the *CMOS* (complementary metal-oxide semiconductor) type. Do not substitute *TTL* (transistor-transistor logic) devices. They cannot handle more than a 6V supply voltage. If you must use TTL devices, add a 5V voltage regulator between the 6V power source and the TTL circuit. Sometimes you might be able to get away with a simple resistive voltage-divider to drop the 6V down to 5V.

All resistors in the projects are generally assumed to be standard 5 percent $^{1}/_{4}$W carbon units unless otherwise noted. Ten-percent resistors should work fine in most of the circuits presented here, but the projects have not been tested with 10 percent resistors. In most cases you can substitute values other than those specified in the parts lists. Generally, use a resistor with a value as close as possible to the specified value, except in cases where the text encourages experimentation. In some cases, you will want to experiment with substituting a potentiometer (or trimmer potentiometer) for a fixed resistor.

Resistors can be wired together in series, as shown in Fig. 2-5, to create a larger effective resistance. Calculate the total effective resistance simply by adding the individual series values:

$$RT = R1 + R2 + R3 \ldots + Rn$$

As an example, assume you have three 180Ω (ohm) resistors wired in series. In this case, the total resistance (ignoring the resistor tolerances) will be equal to:

$$\begin{aligned} RT &= 180 + 180 + 180 \\ &= 540\Omega \end{aligned}$$

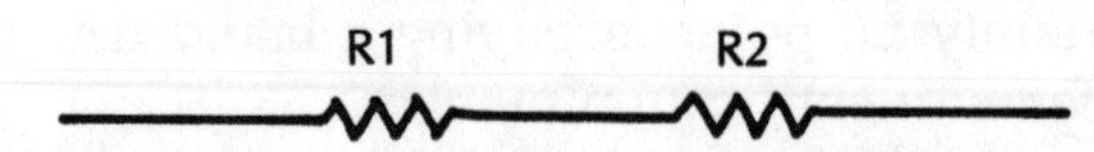


***Fig. 2-5*** *The equivalent value for resistances in series is the sum of the individual component resistances.*

Notice that the total effective resistance for any series combination of resistors is always larger than any of the individual series resistances.

Resistors can also be combined in parallel, as illustrated in Fig. 2-6. The equation for the total effective resistance is slightly more complex for parallel resistances. The reciprocal of the total effective resistance is equal to the sum of the reciprocals of the individual parallel resistances. This sounds very complicated, but it is a lot simpler if you put it in the form of an equation:

$$1/RT = 1/R1 + 1/R2 + 1/R3 \ldots + 1/Rn$$

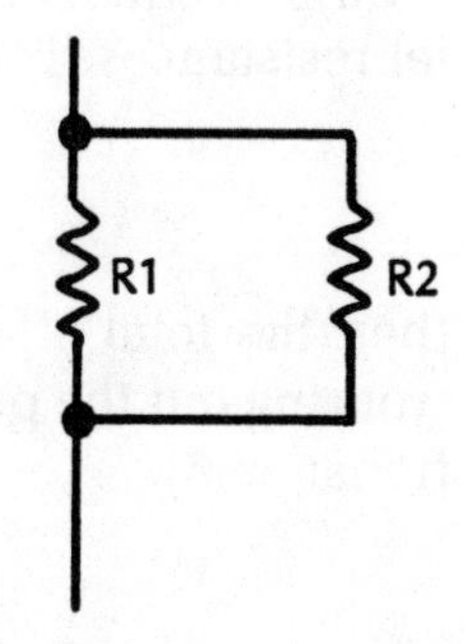


***Fig. 2-6*** *The equivalent value for resistances in parallel is always less than any of the individual component resistances.*

For example, if three 100Ω resistors are combined in parallel, the total effective resistance (again ignoring the effects of the resistor tolerances), will work out to:

$$\begin{aligned} 1/RT &= 1/100 + 1/100 + 1/100 \\ &= 3/100 \\ &= 1/33.3 \\ &= 33.3\Omega \end{aligned}$$

Notice that the total effective resistance of a parallel resistor combination is always less than the value of any of the individual parallel resistances.

If just two resistances are combined in parallel you can use a slightly different alternate equation:

$$RT = (R1 \times R2)/(R1 + R2)$$

As an example, assume R1 is 100Ω and R2 is 200Ω. In this case, the total effective resistance (ignoring resistor tolerances) will be:

$$\begin{aligned} RT &= (100 \times 220)/(100 + 220) \\ &= 22000/320 \\ &= 68.75\Omega \end{aligned}$$

This alternate equation will give exactly the same results as the general parallel resistance equation, but it is sometimes a bit more convenient to use.

You do need to consider one special case. If both parallel resistances have equal values, the total effective resistance will always be equal to exactly one half the value of either of the parallel resistances. For example, if:

$$R1 = R2 = 100\Omega$$

then the total effective resistance will be 50Ω. Confirm this by working out the problem with one of the parallel resistance equations:

$$\begin{aligned} RT &= (100 \times 100)/(100 + 100) \\ &= 10000/200 \\ &= 100/2 \\ &= 50\Omega \end{aligned}$$

Of course, series and parallel resistance combinations can be used together. A sample series/parallel network is shown in Fig. 2-7. To determine the total effective resistance of such a resistance network, you need to break the circuit down into individual series or parallel combinations and solve for the total effective resistance step by step.

In the sample circuit, first solve for the series value of resistances Ra and Rb. Call this series combination Rab. The network is redrawn in Fig. 2-8 to show this series combination as a single resistance element.

Next, find the parallel value of Rc and Rab. Figure 2-9 shows the resistance network redrawn once more to show this combined resistance (Rabc) as a single component. Now, simply find the series combination of the values for Rd and Rabc. This will be the total effective resistance (RT) for the entire series/parallel resistance network.

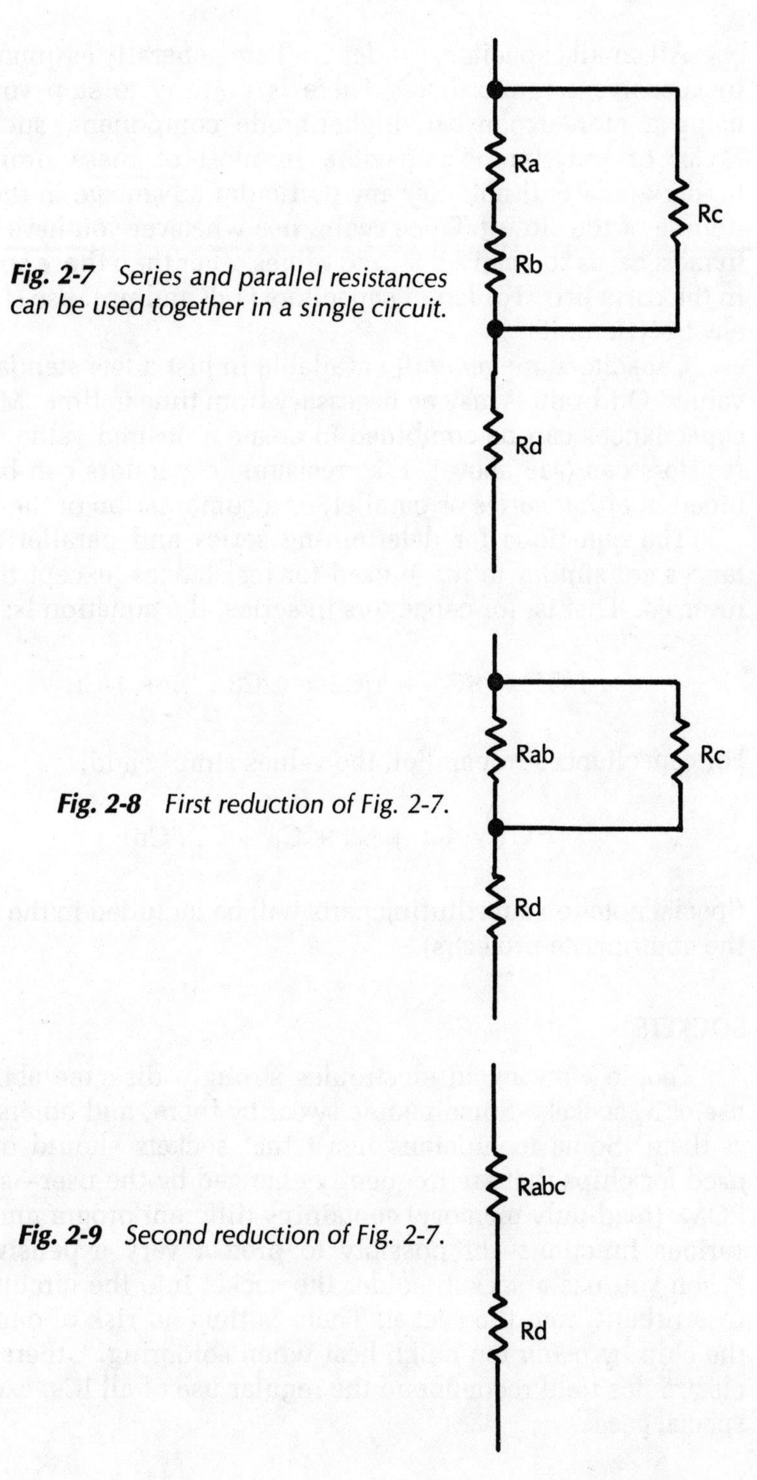


***Fig. 2-7*** *Series and parallel resistances can be used together in a single circuit.*

***Fig. 2-8*** *First reduction of Fig. 2-7.*

***Fig. 2-9*** *Second reduction of Fig. 2-7.*

All small capacitors (under 1$\mu$F) are generally assumed to be inexpensive ceramic discs. There is nothing to stop you from using a more-expensive, higher-grade component, such as a Mylar or polystyrene capacitor. In most of these projects, a higher-grade will not offer any particular advantage in the functioning of the circuit. Once again, use whatever you have handy. In most cases you can substitute values other than those specified in the parts lists. For larger capacitors (1$\mu$F and up), use standard electrolytic units.

Capacitors are generally available in just a few standardized values. Odd values may be necessary from time to time. Multiple capacitances can be combined to create a desired value, just as resistors can (see above). Like resistors, capacitors can be combined in either series or parallel, or a combination of the two.

The equations for determining series and parallel capacitances are similar to those used for resistances, except they are reversed. That is, for capacitors in series, the equation is:

$$1/CT = 1/C1 + 1/C2 + 1/C3 \ldots + 1/Cn$$

For capacitances in parallel, the values simply add;

$$CT = C1 + C2 + C3 + \ldots Cn$$

Special notes on substituting parts will be included in the text for the appropriate project(s).

## SOCKETS

People working in electronics strongly disagree about the use of IC sockets. Some people swear by them, and others swear at them. Some technicians insist that sockets should only be used for chips that are frequently changed by the user—such as ROMs (read-only memory) containing different programming for various functions—or possibly to protect very expensive ICs. When you use a socket, solder the socket into the circuit, then insert the IC into the socket. There is thus no risk of damaging the chip by using too much heat when soldering. Others in the electronics field recommend the regular use of all ICs, except in special cases.

It might seem pretty silly to protect a 25-cent IC with a 50-cent socket, but what you are really protecting is your own time and sanity. If you make a mistake, apply a little too much heat, or if an IC has to be replaced for servicing at a later date, an IC socket can simplify the job immensely. Without a socket, you will have to desolder and then resolder each individual pin on the IC package, while carefully watching out for solder bridges and possible over-heating. That process is not worth the trouble, especially if such problems can so easily be avoided. Sockets do not add that much to the cost of a project, and they can head off a lot of grief and frustration if problems arise. Think of IC sockets as a sort of insurance policy.

Those who oppose sockets often point out how easy it is to insert an IC backward into a socket. When power is applied to the circuit, a backward IC will almost certainly be damaged because the power supply voltages will be fed to the wrong pins. However, it is just as easy to install an IC backward even if a socket is not used. Either you work carefully and make it a practice of double checking before applying power, or you do not. It is not the fault of the socket.

There are some special cases in which the use of IC sockets will be undesirable. In equipment intended for field use, sockets might not be advisable. Such equipment is likely to bounce around a lot. Direct soldering might be a good idea in such equipment to prevent a chip pin from bouncing out of place.

A few (very few) high-frequency circuits can be disturbed by the slightly poorer electrical connections resulting from the use of a socket. But these problem circuits are few and far between. In 99 percent of the circuits you are likely to work with (including all of those presented in this book), using IC sockets will cause no problems and could save you from a lot of needless hassle.

## HEAT SINKS

Where semiconductor components (such as transistors or integrated circuits) handle moderate to large amounts of power, a heat sink will generally be necessary to dissipate heat generated within the semiconductor device itself. A transistor or IC, if unprotected, could literally self-destruct if it tries to pass too much current with no way for the heat generated to be carried away from the delicate semiconductor crystal.

A *heat sink* is nothing more than some kind of heat conductor designed to carry heat away from a temperature-sensitive device and dissipate it, usually into the surrounding air.

Most heat sinks are metallic shields (usually with fins to maximize the area exposed to the surrounding air) that are fitted over the component to be protected. Metal heat sinks of various shapes and sizes are available from many sources of electronics components. When in doubt, use the next larger size. It is better to have too much heat sinking than too little.

Heat transfer between the component and the actual heat sink can be maximized by using a special heat sink compound between the two. Heat sink compound is available from Radio Shack stores and many other electronics suppliers.

On some printed circuit boards, large copper pads can be included to serve as simple heat sinks, if large amounts of power are not involved. ICs that are suitable for heat sinks of this type (including a number of amplifier ICs) are often fitted (by the manufacturer of the chip) with one or more tabs for soldering directly to the copper pads forming the heat sink.

You can calculate the area of copper (size of the pad) needed for a heat sink quite simply if you know the relevant circuit parameters. First, it is necessary to determine the maximum power that heat sink must dissipate. You can use this equation:

$$\text{Power (in watts)} = 0.4 \times (V_s/8R_L) + (V_s \times I_d)$$

where $V_s$ is the maximum supply voltage, $I_d$ is the quiescent drain current (in amperes) under the most adverse conditions (worst-case figure), and $R_L$ is the load resistance (for example, the loudspeaker impedance in the case of an audio amplifier circuit).

Strictly speaking, the value of $V_s$ used in this equation should be the power supply voltage plus an extra 10 percent. For example, if the circuit is powered by a 6V battery, the value of $V_s$ for use in the equation is:

$$\begin{aligned} V_s &= 6 + (10 \text{ percent of } 6) \\ &= 6 + (0.1 \times 6) \\ &= 6 + 0.6 \\ &= 6.6\text{V} \end{aligned}$$

This extra 10 percent allows for any possible fluctuations in the power level. For instance, a very fresh battery might put out more than its rated voltage. If the circuit has a stabilized power supply, such as a voltage regulator, then $V_s$ can be taken simply as the regulated supply voltage.

The quiescent drain current ($I_d$) is usually found among the parameters in the manufacturer's specification sheet for the chip. This value will depend on the supply voltage used in the circuit. On most IC specification sheets, $I_d$ figures are often quoted for typical and maximum. In this case, use the maximum value in the equation.

There is no real reason the pad has to be in a square shape. This is just the simplest shape for determining the area. If there is any margin of error, always try to overestimate the area. The calculated area is the minimum size for the required amount of heat sinking. The copper-pad method is only suitable for low power dissipation. If a great deal of heat must be dissipated, you must use a metal heat sink.

## CUSTOMIZING THE PROJECTS

As you read about the projects in this book, you are likely to feel that some may come close to your individual needs, but are not quite right. Feel free to customize any or all of the projects to suit your own individual applications.

Often you can ignore the original intended application (as described in the text) and try to determine if the circuit can be put to work for the application you have in mind. Use your imagination. Frequently a small change in a circuit will result in a totally different device. For example, a circuit that responds to changes in lighting levels could be made to respond to changes in temperature, perhaps by just substituting a thermistor for the original photoresistor. In many cases, considerable customization can be achieved simply by changing a sensor or by changing the input signal. Sometimes two or more simple projects can be combined to create a moderate to complex system.

Block diagrams can be a big help when designing customized systems. Determine what each stage in the circuit needs to do, then find a circuit that serves that function. This is far easier than designing a complex circuit from scratch.

Actually, most (if not all) complex circuits built by either commercial manufacturers or hobbyists can be broken down into several relatively simple stages, or subcircuits. As always, use your imagination and take your projects as far as you can.

Breadboard any circuit changes before permanently soldering them. Occasionally, as many hobbyists and technicians have learned to their grief, what works on paper may not work in quite the same way in an actual circuit. It is better to find any problems or surprises early on, when it is easy to make additional changes and re-use the components. If you wait until you have got the whole circuit soldered together before you test out your modifications, you are putting out an open invitation to frustration.

# ❖3 LED (Light-Emitting Diode) Flasher Projects

FOR SOME REASON, LIGHT FLASHERS ARE POPULAR SIMPLE PROJECTS among electronics hobbyists. Such projects might be considered frivolous by some. Well, so what? Who says technology cannot be fun?

An LED flasher is nothing more than a circuit that turns one or more LED on and off in a regular pattern. Imaginative people can come up with practical uses. Some obvious applications include warning signals or eye-catching displays for advertising or other purposes. For the most part, however, LED flasher projects are built just for the fun of it. This type of project is often called a *do-nothing* box, because its activity (blinking the light on and off) does not really accomplish anything.

There is something inherently fascinating and even a little hypnotic about a flashing LED. Every electronics hobbyist should try a few LED flasher circuits. You might be surprised at how easy it is to get hooked.

LED flasher projects are especially popular with beginners in the electronics field. Projects of this sort tend to be quite simple to build, and they do something noticeable. No expensive equipment is needed to calibrate the circuit or to determine if it is working properly. In fact, no equipment at all is needed to check the operation of an LED flasher project, unless you count your own eyes as equipment. The components required are typically quite inexpensive and easy to find.

Flasher circuits have always been popular, even before the LED itself was invented. Earlier circuits blinked neon lamps or flashlight bulbs on and off. LEDs offer several advantages. They do not burn out (or at least, not as fast as an incandescent lamp). LEDs use less power than lamps, and they are considerably smaller and less expensive too. Frankly, LEDs look a lot snazzier than neon lamps or flashlight bulbs.

## PROJECT 3: SIMPLE LED FLASHER

You can use any of many possible circuits to flash an LED on and off. A particularly simple LED flasher circuit is shown in Fig. 3-1, and the parts list is in Table 3-1. This project is built around a pair of CMOS NOR gates. Actually, you could use a pair of NAND gates or inverters. The two inputs of each NOR gate are shorted together, effectively creating an inverter stage. Shorting together the inputs of a NAND gate will have the same effect. NOR gates are called for here simply because the CD4001 quad NOR gate IC is one of the most readily available CMOS chips. You can use any CMOS NOR, NAND, or inverter device you happen to have.

The flash rate for the LED is controlled primarily by resistor R2 and capacitor C1. Basically, this circuit is nothing more than a simple, low-frequency oscillator. The output alternates between HIGH and LOW states in the form of a square wave. When the output signal is HIGH, the LED is lit; when the output state is LOW, the LED is turned off. The oscillation effect is caused by the feedback path between the two inverter stages (IC1A and IC1B).

Reducing the value of either resistor R2 or capacitor C1 (or both) will increase the frequency of the oscillation. The LED will

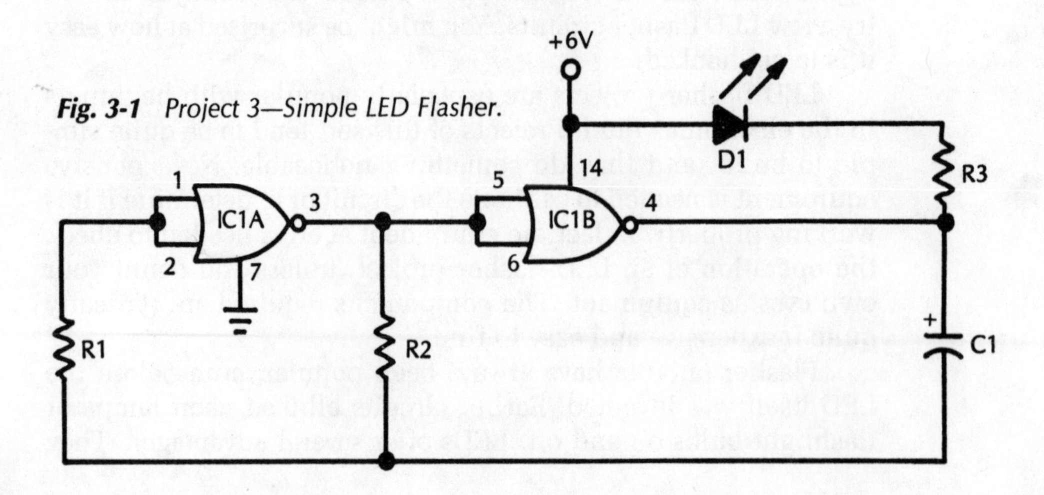


*Fig. 3-1* Project 3—Simple LED Flasher.

**Table 3-1 Parts List for Project 3—Simple LED Flasher.**

| Part | Component Needed |
|---|---|
| IC1 | CD4001 quad NOR gate |
| D1 | LED |
| C1 | 10μF 15V electrolytic capacitor* |
| R1 | 1MΩ 1/4W resistor |
| R2 | 180kΩ 1/4W resistor* |
| R3 | 330Ω 1/4W resistor |

*Frequency-determining component; see text.

blink on and off at a faster rate. The frequency, or flash rate, is inversely proportional to the values of these two components. Of course, increasing the value of R2 and/or C1 will reduce the frequency, slowing down the flash rate. It will take longer for the LED to blink on and off.

If the frequency is set for too high a level, the LED will appear to be continuously lit. It is still blinking on and off at a very high rate, but your eyes are incapable of distinguishing the individual blinks. This is the same principle used in motion picture film. (The individual images change so fast that they appear to be in continuous motion.) For a functional LED flasher, the oscillator frequency should be no higher than about 6 to 10 *Hz* (Hertz); that is, the LED should not blink on and off more than 6 to 10 times per second. For best results, use slower flash rates.

If you would like to modify this LED flasher circuit for a manually variable flash rate, use a small value potentiometer in series with a relatively large (150kΩ or so) fixed resistor in place of resistor R2 shown in the diagram.

If you experiment with alternate values of timing resistor R2, try to give resistor R1 a value of about 5 to 10 times that of R2. The exact value of R1 is not at all critical, but it should be significantly larger than R2 to ensure circuit stability.

There is just one more component in this circuit: resistor R3. This is a current-limiting resistor to protect the LED (D1). Without this resistor, the LED will draw too much current and quickly self-destruct. Changing the value of the current-limiting resistor will affect the brightness of the LED when it is lit. The smaller the resistance, the brighter the LED will glow. Resistor R3 should have a value in the 200 to 1kΩ range. Experiment with other values for the timing components (resistor R2 and capacitor C1).

## PROJECT 4: ALTERNATE LED FLASHER

Another simple LED flasher circuit is illustrated in Fig. 3-2. A typical parts list for this project appears as Table 3-2. Once again, this circuit is basically a low-frequency oscillator. The output that switches between LOW and HIGH times, which turns the LED on and off. Instead of being built around a pair of inverter stages, as in project 3, this oscillator circuit is built around the popular 555 timer IC. You can substitute the CMOS version of this chip (7555) if you prefer.

In project 3, the oscillator put out a true square wave. The output is HIGH for exactly half of each cycle, and LOW for the other half. The LED on and off times will be identical. (There might be some minor differences in the times, but they will be so small that they are negligible. You would find it difficult to even detect any differences in the timing.)

The 555 oscillator circuit used in this project puts out a rectangle wave. The ratio of on and off times is not 50 percent (or 1 to 2). The ratio of on and off times in a rectangular wave is called

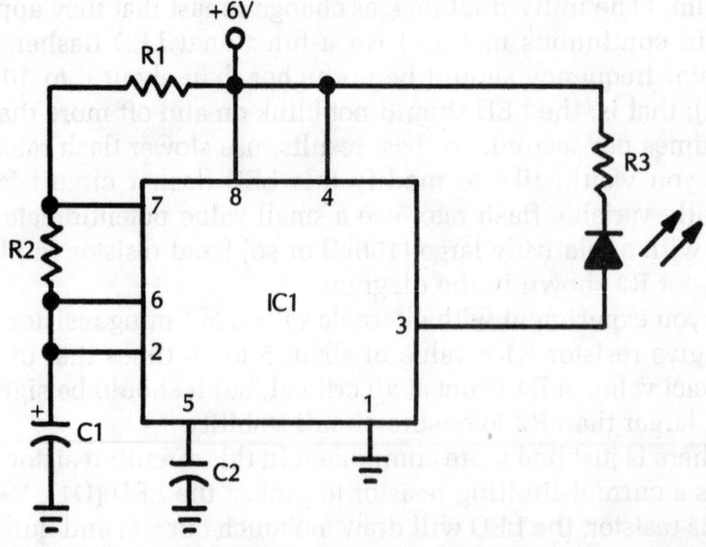


*Fig. 3-2* *Project 4—Alternate LED Flasher.*

**Table 3-2 Parts List for Project 4—Alternate LED Flasher.**

| Part | Component Needed |
|---|---|
| IC1 | 555 timer (or 7555) |
| D1 | LED |
| C1 | 10μF 15V electrolytic capacitor* |
| C2 | 0.01μF capacitor |
| R1 | 22kΩ 1/4W resistor* |
| R2 | 10kΩ 1/4W resistor* |
| R3 | 330Ω 1/4W resistor |

*Experiment with other component values.

the *duty cycle.* The duty cycle in this circuit will be determined by the relative values of resistors R1 and R2.

Three components determine the timing in this circuit. They are resistors R1 and R2 and capacitor C1. The output frequency is determined by the following equation:

$$F = 1/(0.693\ C1\ (R1 + 2R2))$$

This looks a bit complicated, but you can break it down into simpler steps.

The output switches back and forth between the HIGH and LOW states. The output HIGH time is different from the output LOW time, so obviously there must be different equations for these two durations.

For the output LOW time, during which the LED will be dark (off), only the values of resistor R2 and capacitor C1 are of significance. Resistor R1 is ignored by the circuit for this portion of the cycle. The equation is:

$$T_L = 0.693C1R2$$

Substitute the component values from the part list (Table 3-2), to find that the output will be LOW (and the LED will be off) for a period of:

$$C1 = 10\ \mu F$$
$$R2 = 100K$$

$$\begin{aligned} T_L &= 0.693 \times 0.00001 \times 100{,}000 \\ &= 0.693 \text{ second} \\ &\cong 0.69 \text{ second} \end{aligned}$$

To determine how long each cycle the output is HIGH (and the LED is on), consider the value of resistor R1:

$$T_H = 0.693\ C1\ (R1 + R2)$$

Substituting the component values from the parts list, find the nominal on time for our project:

$$\begin{aligned} R1 &= 22k\Omega \\ R2 &= 100k\Omega \\ C1 &= 10\mu F \end{aligned}$$

$$\begin{aligned} T_H &= 0.693 \times 0.00001 \times (22000 + 100000) \\ &= 0.693 \times 0.00001 \times 122000 \\ &= 0.74546 \text{ second} \\ &\cong 0.75 \text{ second} \end{aligned}$$

You can find the total cycle time simply by adding the LOW and HIGH times:

$$T_T = T_L + T_H$$

For the Fig. 3-2 circuit:

$$\begin{aligned} T_T &= 0.69 + 0.75 \\ &= 1.44 \text{ second} \end{aligned}$$

The circuit goes through one complete cycle every 1.44 seconds. The frequency is simply the reciprocal of the cycle time. That is:

$$F = 1/T_T$$

$$\begin{aligned} F &= 1/1.44 \\ &\cong 0.0069 \text{ Hz} \end{aligned}$$

Experiment with other values for capacitor C1, resistor R1, and resistor R2 to achieve different output timings.

Include capacitor C2 to stabilize the timer. It is not necessary in all cases, but it is cheap insurance. There is nothing to be gained by experimenting with alternate values for this capacitor because it does not directly affect circuit operation.

## PROJECT 5: DUAL LED FLASHER

A simple LED flasher—like the circuits of project 3 and project 4—is fun, but it can quickly become boring. There is just a single LED flashing on and off. Multiple flashing LEDs can offer far more variety. A dual LED flasher circuit is illustrated in Fig. 3-3, and the parts list is in Table 3-3. This project is just a variation of project 3 (simple LED flasher). The original circuit is shown in Fig. 3-1.

Project 5 adds a second LED and an extra inverter stage to the basic LED flasher circuit. The two LEDs will always see opposite-state signals. When LED1 is on, LED2 will be off, and vice versa. As long as power is supplied to the circuit, one (and only one) of the LEDs will be lit. They will never be simultaneously dark (unless the supply voltage is removed), and they will never be simultaneously lit. If a high frequency is being put out by the oscillator, both LEDs might appear to be continuously lit, but they are still blinking on and off alternately, at a rate too fast for you to see.

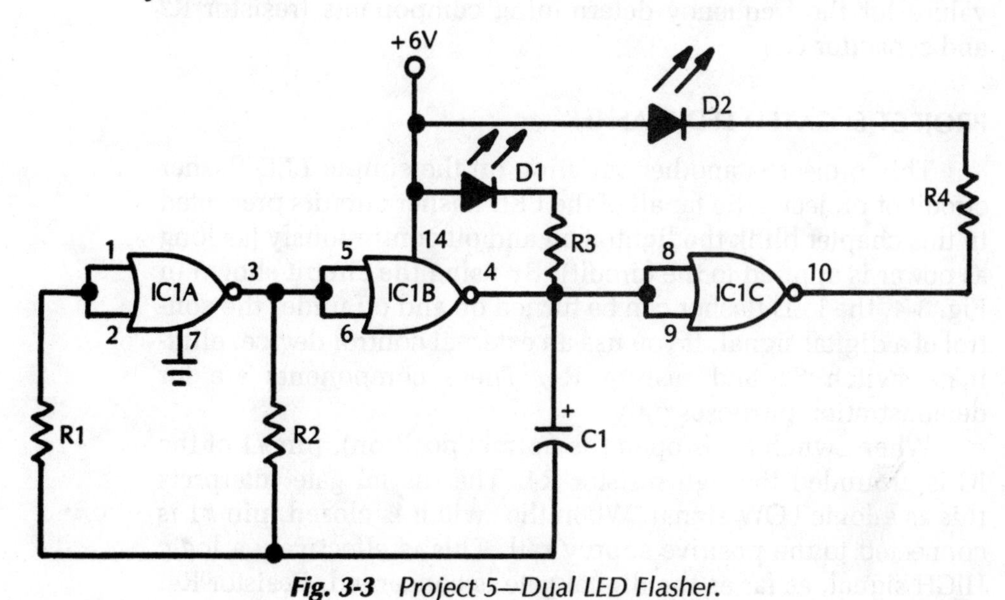


*Fig. 3-3* *Project 5—Dual LED Flasher.*

**Table 3-3 Parts List for Project 5—Dual LED Flasher.**

| Part | Component Needed |
|---|---|
| IC1 | CD4001 quad NOR gate |
| D1, D2 | LED |
| C1 | 10μF 15V electrolytic capacitor |
| R1 | 1MΩ 1/4W resistor |
| R2 | 180kΩ 1/4W resistor |
| R3, R4 | 330Ω 1/4W resistor |

The basic LED flasher project (project 3) uses two stages of a CD4001 quad NOR gate. There are still two unused stages. One of these unused stages can be used as the added third inverter stage. The only new components required for this modification of the basic circuit are the second LED (D2), and its current-dropping resistor (R4). These components simply duplicate the operation of D1 and R3 in project 3.

Notice that the parts list (Table 3-3) is the same as the parts list in Table 3-1, except for the addition of D2 and R4. As with the basic LED flasher project, experiment with other component values for the frequency determining components (resistor R2 and capacitor C1).

## PROJECT 6: GATED LED FLASHER

This project is another variation on the simple LED flasher circuit of project 3. So far all of the LED flasher circuits presented in this chapter blink the light(s) on and off continuously (as long as power is applied to the circuit). By using the circuit shown in Fig. 3-4, the LED flasher can be turned on and off under the control of a digital signal. If you use an external control device, eliminate switch S1 and resistor R4. These components are for demonstration purposes only.

When switch S1 is open (its normal position), pin #1 of the IC is grounded through resistor R4. The digital gate interprets this as a logic LOW signal. When the switch is closed, pin #1 is connected to the positive supply rail. This is effectively a logic HIGH signal, as far as the digital gate is concerned. Resistor R4,

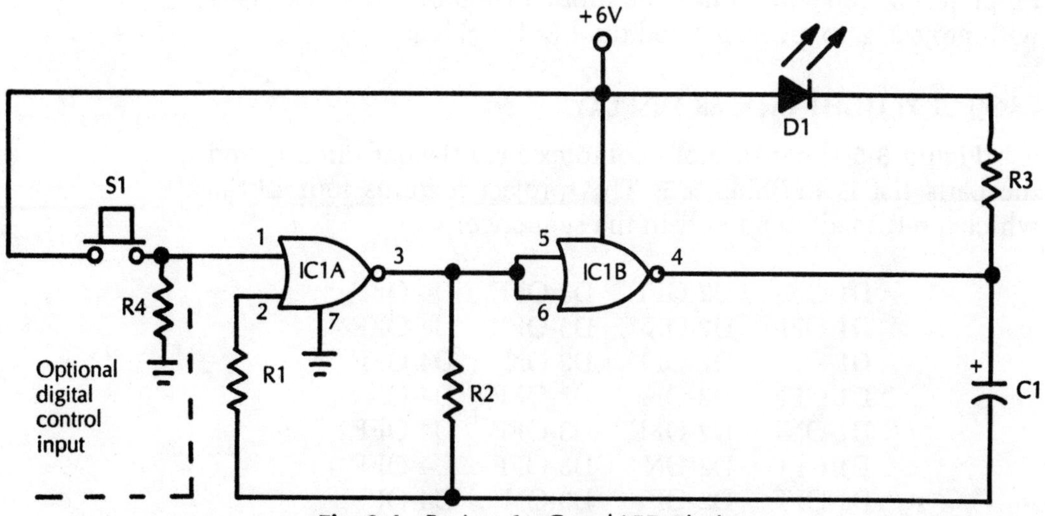


*Fig. 3-4* *Project 6—Gated LED Flasher.*

**Table 3-4 Parts List for Project 6—Gated LED Flasher.**

| Part | Component Needed |
|---|---|
| IC1 | CD4001 quad NOR gate |
| D1 | LED |
| C1 | 10μF 15V electrolytic capacitor |
| R1 | 1MΩ 1/4W resistor |
| R2 | 180kΩ 1/4W resistor |
| R3 | 330Ω 1/4W resistor |
| R4 | 1MΩ 1/4W resistor* |
| S1 | *SPST* (single-pole, single-throw) normally open push-button switch* |

*See text.

which has a large value, prevents switch S1 from shorting the power supply voltage directly to ground. The exact value of resistor R4 is not at all critical, so long as it is large.

The parts list for this gated LED flasher circuit is given in Table 3-4. Switch S1 and resistor R4 are optional. They are eliminated if an external source of digital control signals is used. Otherwise, the parts list for this project is exactly the same as the one for project 3. You should have no problem combining this project with project 4, to create a gated dual LED flasher.

## PROJECT 7: LIGHT-CHASER DISPLAY

Figure 3-5 shows a more complex LED flasher circuit, and the parts list is in Table 3-5. This project features four LEDs, which are turned on and off in the sequence:

| | | | |
|---|---|---|---|
| D1-ON | D2-OFF | D3-OFF | D4-OFF |
| D1-OFF | D2-ON | D3-OFF | D4-OFF |
| D1-OFF | D2-OFF | D3-ON | D4-OFF |
| D1-OFF | D2-OFF | D3-OFF | D4-ON |
| D1-ON | D2-OFF | D3-OFF | D4-OFF |
| D1-OFF | D2-ON | D3-OFF | D4-OFF |
| D1-OFF | D2-OFF | D3-ON | D4-OFF |
| D1-OFF | D2-OFF | D3-OFF | D4-ON |

and so forth.

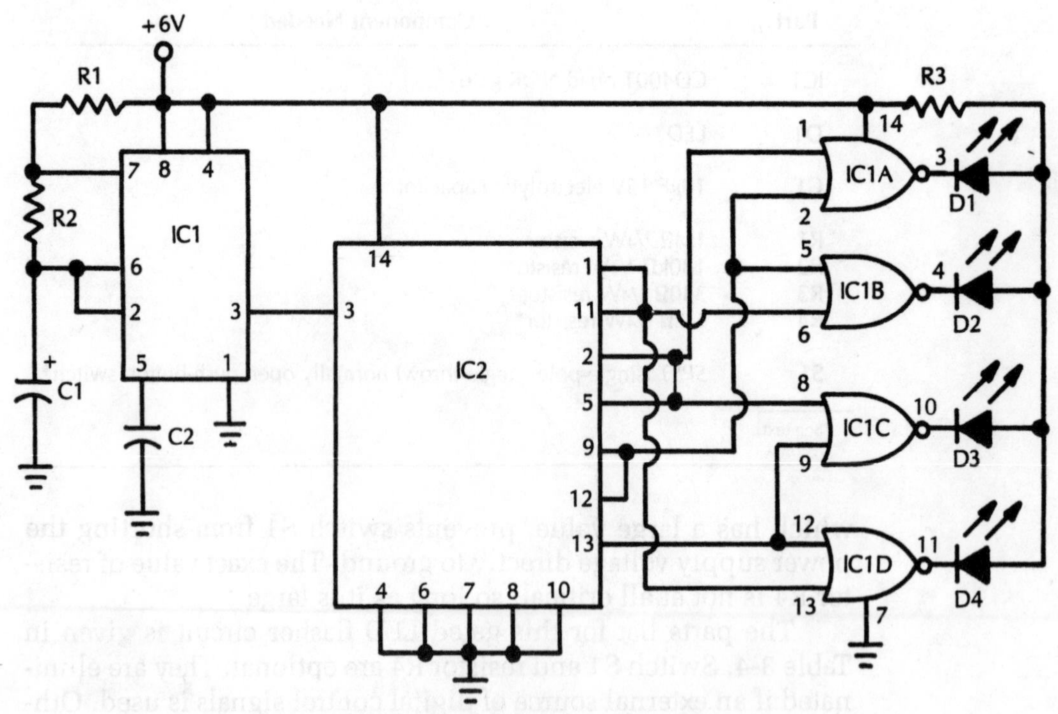


*Fig. 3-5 Project 7—Light-Chaser Display.*

**Table 3-5 Parts List for Project 7—Light-Chaser Display.**

| Part | Component Needed |
|---|---|
| IC1 | 555 timer |
| IC2 | CD4013 dual D flip-flop |
| IC3 | CD4001 quad NOR gate |
| D1 – D4 | LED |
| C1 | 10μF 15V electrolytic capacitor |
| C2 | 0.01μF capacitor |
| R1 | 33kΩ 1/4W resistor |
| R2 | 1kΩ 1/4W resistor |
| R3 | 330Ω 1/4W resistor |

Notice that only one LED is ever lit at any given instant. As long as power is applied to the circuit, one (and only one) of the LEDs will be lit.

Use an oscillator, built around a 555 timer (IC1), as the system clock. If you prefer, use the CMOS 7555 timer instead of the more common 555.

Modify the clock frequency by changing the values of capacitor C1 and resistors R1 and R2. Feel free to experiment with these component values. Of course, as with all LED flashers, if the clock frequency is set too high, your eyes will blend the individual blinks together, and all of the LEDs will appear to be continuously lit.

IC2 (a CD4013 dual D flip-flop) is wired as a four-stage counter. Each cycle from the oscillator (IC1) triggers the counter (IC2), causing it to advance its count value.

The four NOR gates of IC3 (CD4001) decode the counter outputs to turn on the appropriate LED for each of the four possible count values. Because only one LED will ever be lit at any given instant, you need only a single current-dropping resistor (R3).

Changing the value of resistor R3 will alter the brightness of the LEDs when lit. Changing the value(s) of C1, R1, and/or R2 will affect the frequency or speed of the sequence.

If you arrange the four LEDs in a suitable pattern, you will achieve a light-chasing effect. A single light will appear to move around the pattern. This makes a very effective display.

## PROJECT 8: ELECTRONIC CHRISTMAS TREE

This project is a delightful and unusual ornament. Twenty LEDs blink on and off at varying rates. The circuit is shown in Fig. 3-6, and the parts list is in Table 3-6. By arranging the LEDs in a triangular pattern, as shown in Fig. 3-7, and painting the LED backing (case or circuit board) appropriately, you achieve the effect of a Christmas tree. Of course, you can use other patterns. Use your imagination.

The heart of this circuit is an exciting IC known as the 2240 programmable timer. Unfortunately, this chip is becoming increasingly hard to find. It might not still be in production. Check the surplus houses. This device is an electronic hobbyist's goldmine. If you cannot find a 2240, do not despair, an alternate (but more complex) version of this electronic Christmas tree project will be presented as project 9.

Because the 2240 is not as familiar and common as most of the other ICs used in this book, a pin-out diagram of this device is shown in Fig. 3-8. The internal operation of the 2240 programmable timer is illustrated in Fig. 3-9. Basically, it is a timer (similar to the popular 555) followed by an eight-stage binary counter. Each successive counter output increases the time period by a factor of two as follows:

| PIN # | OUTPUT TIME |
|---|---|
| 1 | 1T |
| 2 | 2T |
| 3 | 4T |
| 4 | 8T |
| 5 | 16T |
| 6 | 32T |
| 7 | 64T |
| 8 | 128T |

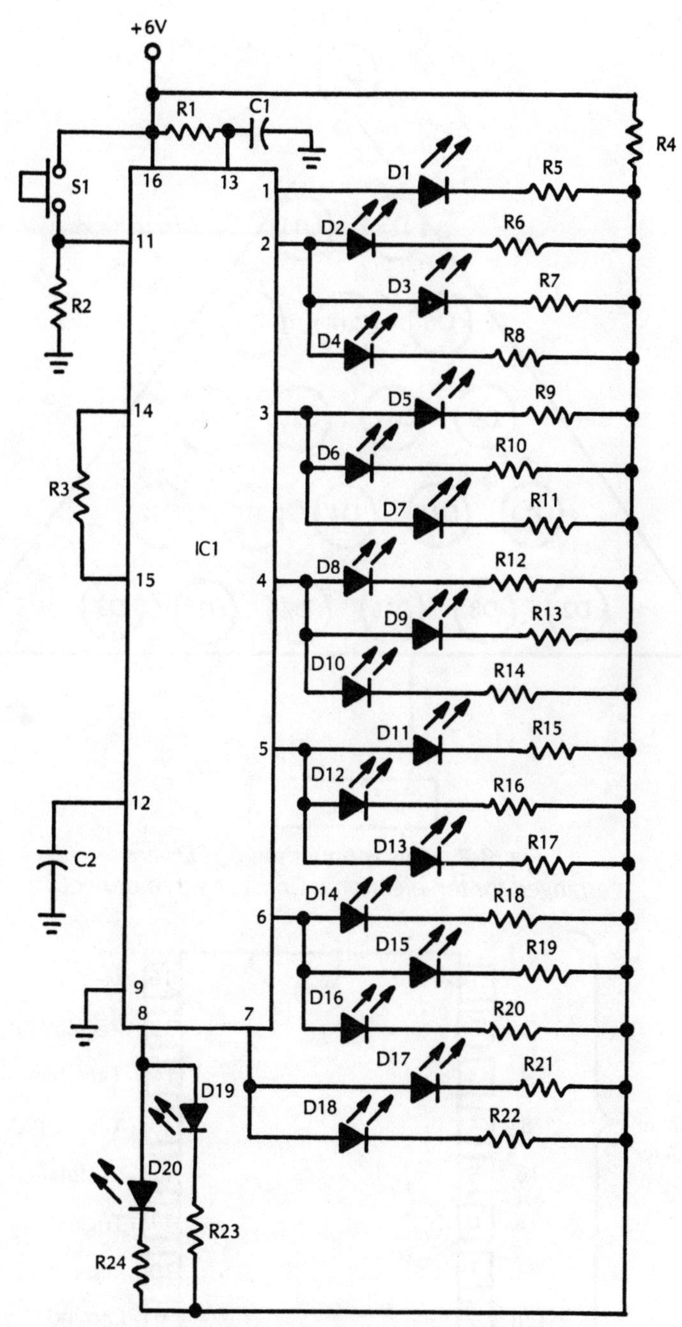


*Fig. 3-6* Project 8—Electronic Christmas Tree.

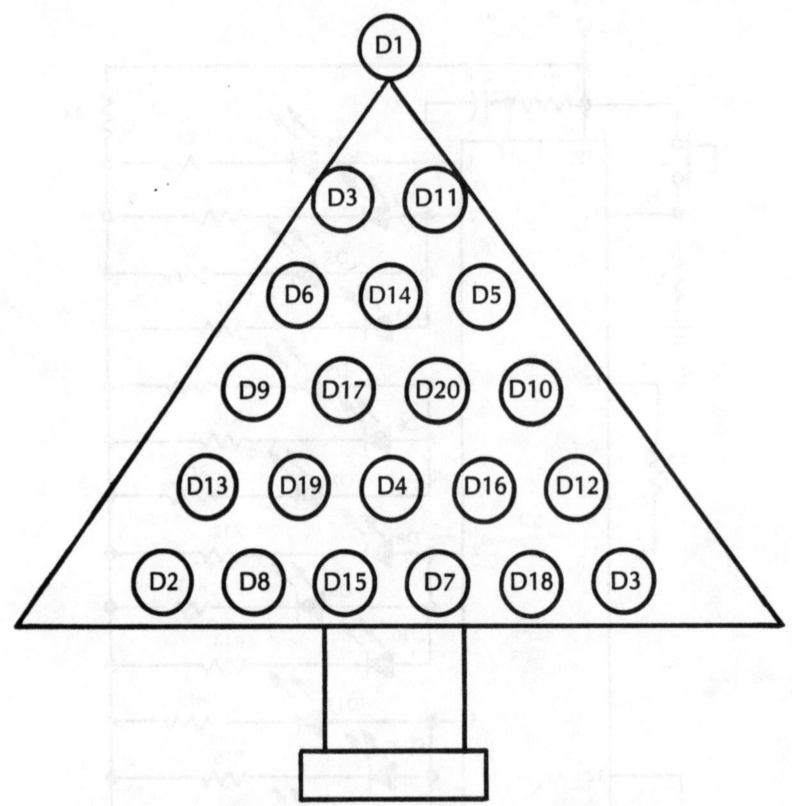


***Fig. 3-7*** *This is how the 20 LEDs are arranged for the electronic Christmas tree project.*

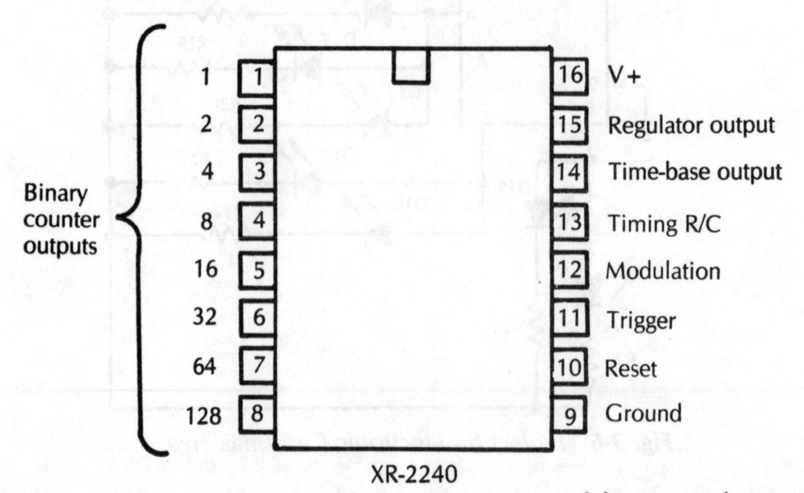


***Fig. 3-8*** *The 2240 programmable timer is a very useful integrated circuit.*

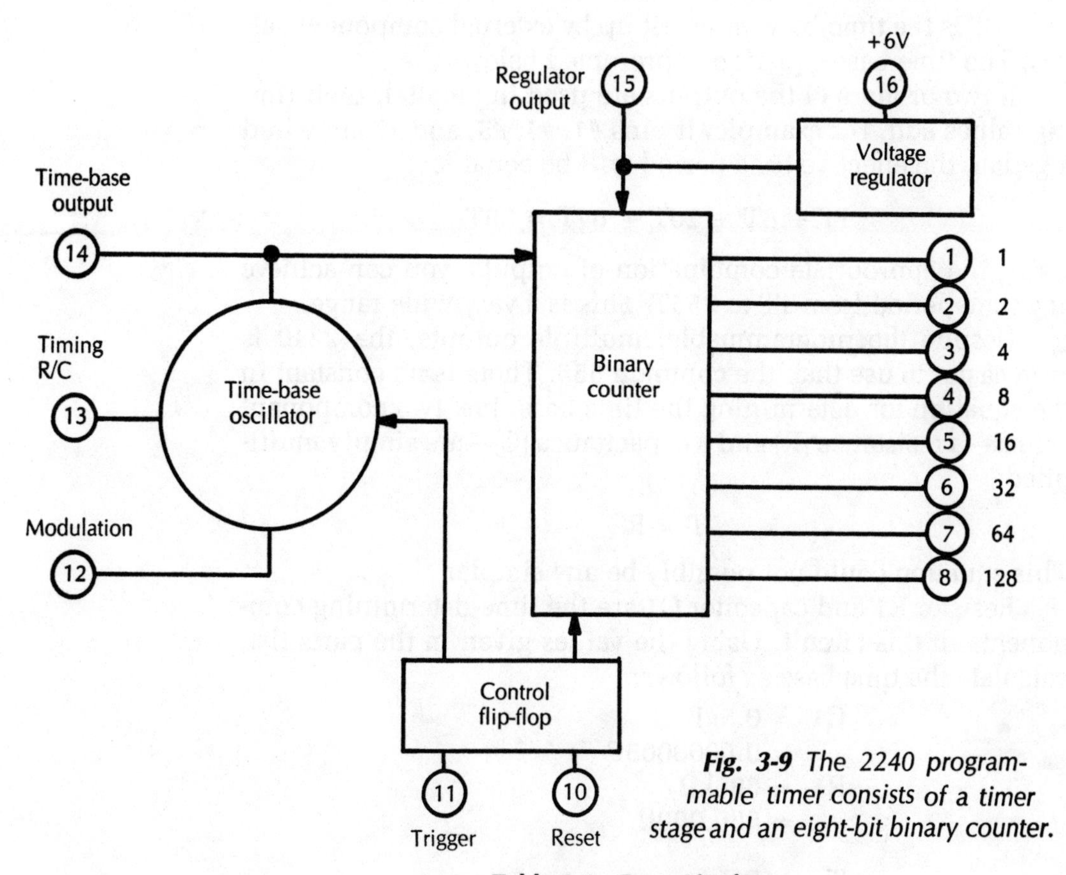


***Fig. 3-9*** *The 2240 programmable timer consists of a timer stage and an eight-bit binary counter.*

**Table 3-6 Parts List for Project 8—Electronic Christmas Tree.**

| Part | Component Needed |
|---|---|
| IC1 | XR2240 programmable timer; see text. |
| D1–D20 | LED |
| C1 | 0.5μF capacitor* |
| C2 | 0.01μF capacitor |
| R1 | 680kΩ 1/4W resistor* |
| R2 | 1MΩ 1/4W resistor |
| R3 | 22kΩ 1/4W resistor |
| R4 | 10kΩ 1/4W resistor |
| R5–R24 | 330Ω 1/4W resistor |

*Frequency determining components; see text.

where T is the time-base value set up by external component values. The time-base equation is presented below.

If two or more of the outputs are used in parallel, their timing values add. For example, if pins #1, #4, #5, and #7 are wired together, the effective time period will be equal to:

$$1T + 8T + 16T + 64T = 89T$$

Using the appropriate combination of outputs, you can achieve any time period from 1T to 255T. This is a very wide range.

Despite the programmable, multiple outputs, the 2240 is even easier to use than the common 555. There is no constant in the equation for determining the time base. Just two component values—a resistance (R) and a capacitance (C)—are simply multiplied:

$$T = RC$$

This equation could not possibly be any simpler.

Resistor R1 and capacitor C1 are the time-determining components in this circuit. Using the values given in the parts list, calculate the time base as follows:

$$\begin{aligned} C1 &= 0.5\mu F \\ &= 0.0000005F \\ R1 &= 680k\Omega \\ &= 680{,}000\Omega \end{aligned}$$

$$\begin{aligned} T &= RC \\ &= 680000 \times 0.0000005 \\ &= 0.34 \text{ second} \end{aligned}$$

An LED connected to pin #1 (1T) will flash on and off about three times a second. Each successive output pin will divide the frequency by a factor of two:

| PIN # | TIME | | FREQUENCY |
|---|---|---|---|
| 1 | 1T | 0.32 | 2.94 Hz |
| 2 | 2T | 0.64 | 1.47 Hz |
| 3 | 4T | 1.28 | 0.74 Hz |
| 4 | 8T | 2.56 | 0.37 Hz |
| 5 | 16T | 5.12 | 0.18 Hz |
| 6 | 32T | 10.24 | 0.092 Hz |
| 7 | 64T | 20.48 | 0.046 Hz |
| 8 | 128T | 40.96 | 0.023 Hz |

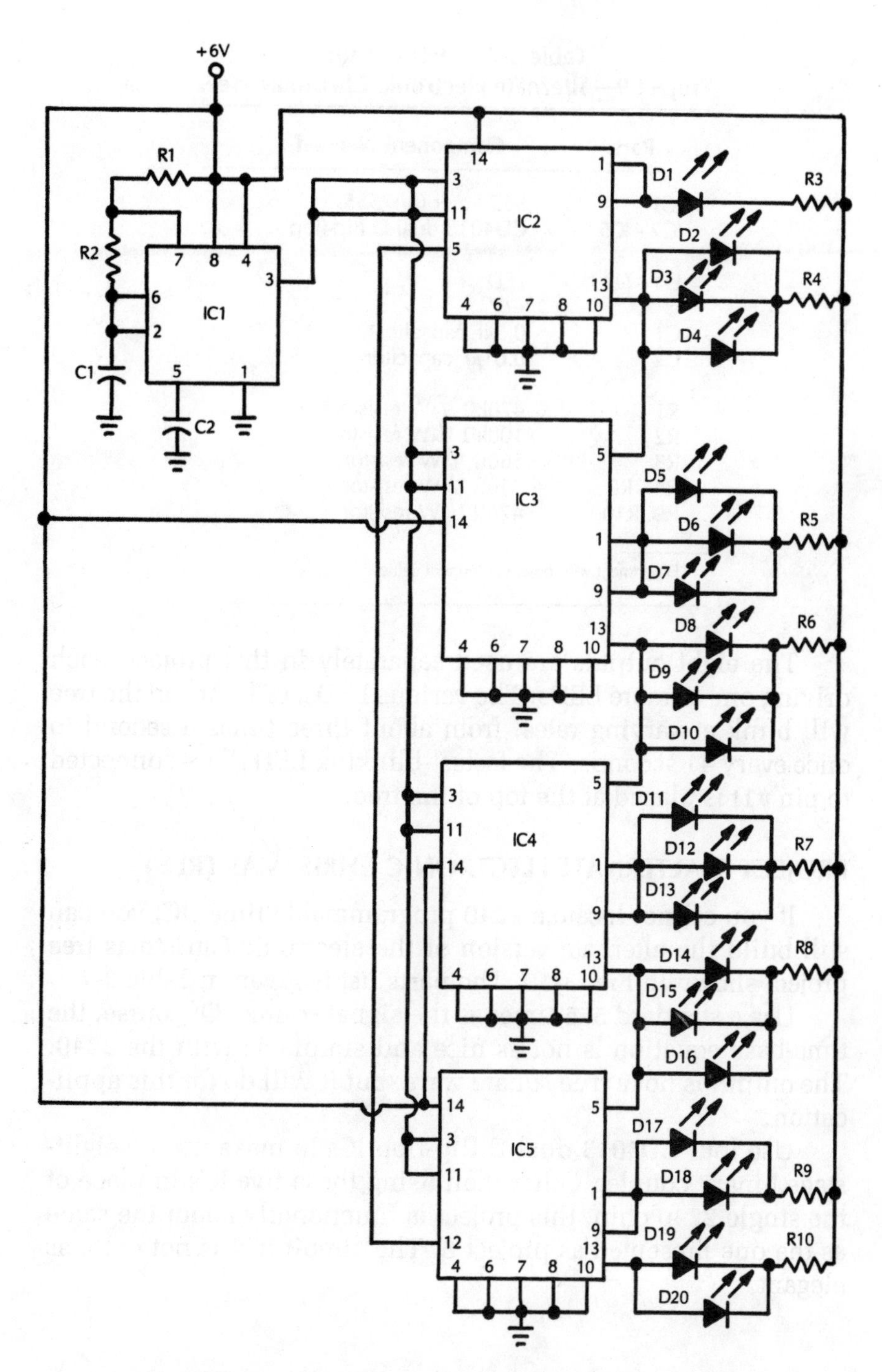


***Fig. 3-10.****Project 9—Alternate Electronic Christmas Tree.*

**Table 3-7 Parts List for Project 9—Alternate Electronic Christmas Tree.**

| Part | Component Needed |
|---|---|
| IC1 | 555 timer (or 7555) |
| IC2 – IC5 | CD4013 dual D flip-flop |
| D1 – D20 | LED |
| C1 | 0.5μF capacitor* |
| C2 | 0.01μF capacitor |
| R1 | 470kΩ 1/4W resistor* |
| R2 | 100kΩ 1/4W resistor* |
| R3 | 560Ω 1/4W resistor |
| R4 – R8 | 330Ω 1/4W resistor |
| R9, R10 | 470Ω 1/4W resistor |

*Experiment with other component values.

The eight outputs are used separately in this project, each driving one or more LEDs. The various LEDs, or lights on the tree will blink at varying rates, from about three times a second to once every 41 seconds. The fastest-blinking LED (D1—connected to pin #1) is placed at the top of the tree.

## PROJECT 9: ALTERNATE ELECTRONIC CHRISTMAS TREE

If you cannot locate a 2240 programmable timer IC, you can still build the alternate version of the electronic Christmas tree project shown in Fig. 3-10. The parts list is given in Table 3-7.

Use a standard 555 timer as the signal source. Of course, the time-base equation is not as nice and simple as with the 2240. The output is not a true square wave, but it will do for this application.

Use four CD4013 dual D flip-flop ICs to make up the eight-stage binary counter. Other than using these five ICs in place of the single 2240 chip, this project is functionally about the same as the one presented as project 8. The circuit just is not quite as elegant.

# ❖ 4 Sound-Generating Circuits

MUSIC- AND SOUND-PRODUCING CIRCUITS ARE ALWAYS POPULAR with hobbyists. Many such projects are quite simple and inexpensive. Like the LED flasher projects of the last chapter, they do something directly, which is not true of all electronic circuits. Your ears can tell you if the circuit is working or not.

The projects in this chapter range from simple oscillators up to complex tone generators and tone sequencers. You will enjoy these sound-generating projects.

## PROJECT 10: SIMPLE OSCILLATOR

An *oscillator* is an electronic circuit that generates an ac waveform. If you feed output signal from an oscillator through an amplifier and a speaker, you will hear a tone. (Of course, the oscillator output frequency must be within the audible range, or about 20 Hz to 20 kHz.)

Many different waveforms can be generated by various circuits. The shape of a wave (as displayed on an oscilloscope) has a strong effect on the perceived quality of the sound heard. Some common waveforms are illustrated in Fig. 4-1. A sine wave (Fig. 4-1A) is very pure and piercing. A triangular wave (Fig. 4-1B) is smoother, almost flutelike. A square wave (Fig. 4-1C) is rather reedy sounding. A sawtooth wave (Fig. 4-1D) is a bit harsh and raspy. You can refer to any of many good books on acoustics that explain the harmonic content of these (and other) waveforms.

A very simple practical oscillator circuit is shown in Fig. 4-2, and a typical parts list for this project is in Table 4-1. This oscillator circuit is built around the popular 555 timer IC. This

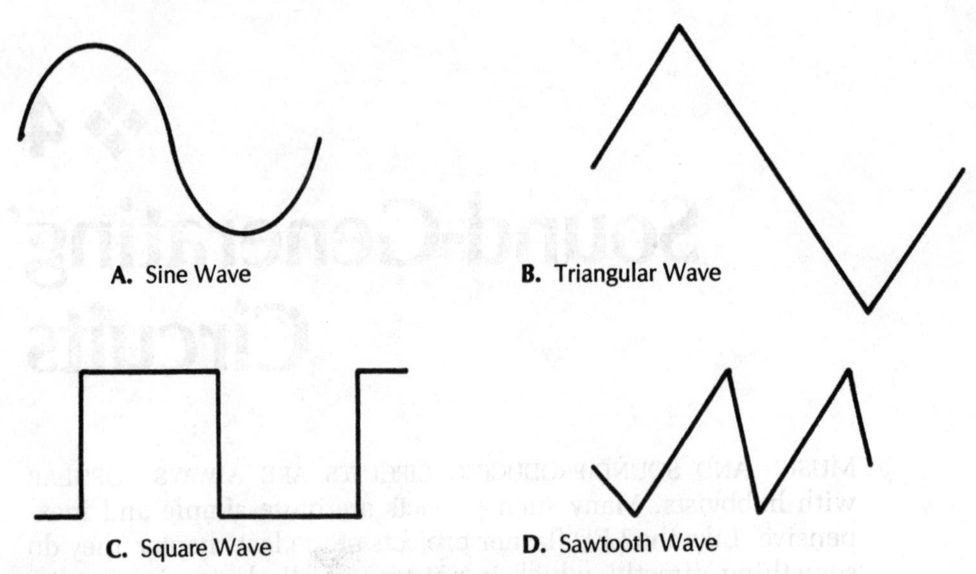


*Fig. 4-1 An oscillator puts out a regular waveform.*

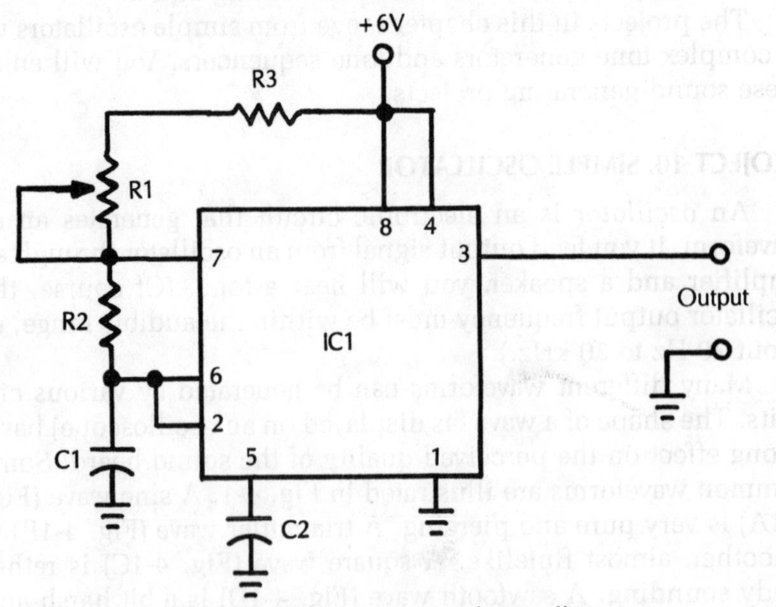


*Fig. 4-2 Project 10—Simple Oscillator.*

chip is cheap, readily available, and easy to work with. The output is in the form of a rectangular wave. The duty cycle is determined by the relative values of resistors R3 and R1 + R2. R1 is a manually variable potentiometer, to give you active control over

**Table 4-1 Parts List for Project 10—Simple Oscillator.**

| Part | Component Needed |
|---|---|
| IC1 | 555 timer |
| C1 | 0.22μF capacitor |
| C2 | 0.01μF capacitor |
| R1 | 10kΩ potentiometer |
| R2 | 1kΩ 1/4W resistor |
| R3 | 2.2kΩ 1/4W resistor |
| Nominal frequency: | 600 Hz |
| Low frequency: | 462 Hz |
| High frequency: | 1562 Hz |

the output frequency. Changing the frequency will alter the duty cycle and thus will alter the tone quality.

A rectangular wave switches back and forth between two definite states: HIGH and LOW. The transition time is nominally instantaneous (actually it is finite, but very short), and can be ignored for practical purposes.

The duty cycle is the ratio of the HIGH time to the total cycle time. A square wave is HIGH for exactly half of each cycle, so the duty cycle is 1 to 2. Because of the design characteristics of the 555, a true square wave is not possible with this circuit, but you still have a wide range of control over the duty cycle.

Use the following equation to find the output frequency of this oscillator circuit:

$$F = 1/(0.693C1(Ra + 2R2))$$

where *Ra* is the sum of *R1* (the variable potentiometer) and *R2* (a fixed resistor). Remember, the duty cycle and the output frequency interact in this circuit.

## PROJECT 11: VCO (VOLTAGE-CONTROLLED OSCILLATOR)

A *VCO* (voltage-controlled oscillator) is at the heart of an electronic music system. VCOs also have many other applications, and they are found in all kinds of equipment.

The component values in a VCO set the nominal output frequency. An input voltage (control voltage) from an external source is fed into the circuit. The output frequency changes proportionately to this control voltage.

The 555 timer IC has a voltage control input at pin #5. Ordinarily, place a small capacitance between this pin and ground for stability. To create a VCO, just delete C2 from the oscillator circuit of project 10 and feed in a control voltage. The VCO circuit is shown in Fig. 4-3, and a typical parts list for this project is in Table 4-2.

In addition to being open to automatic control, this VCO circuit offers another important advantage over the simple oscillator circuit of Fig. 4-2: changing the output frequency by changing the control voltage at the input does not affect the duty cycle of the waveform.

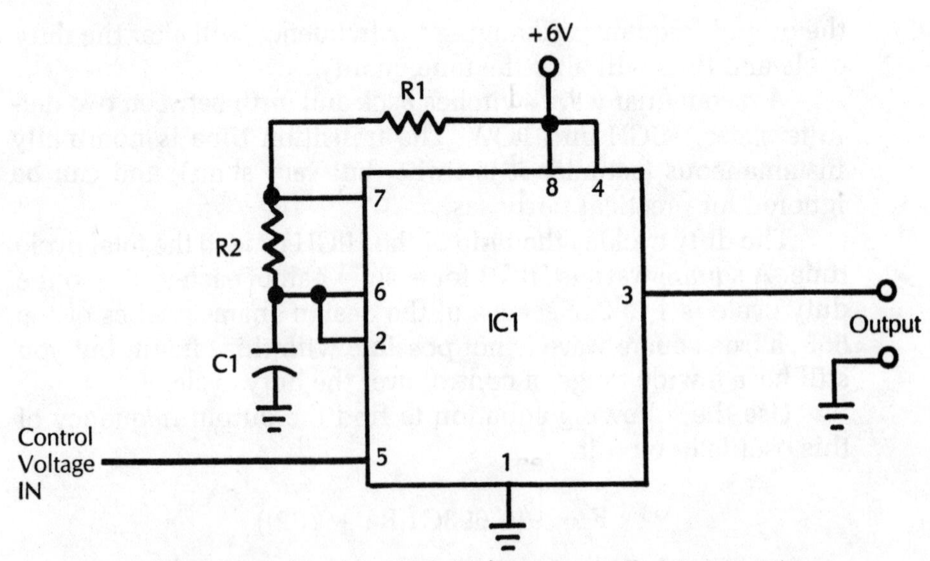


*Fig. 4-3* *Project 11—VCO (Voltage-Controlled Oscillator).*

**Table 4-2 Parts List for Project 11—VCO (Voltage-Controlled Oscillator).**

| Part | Component Needed |
|---|---|
| IC1 | 555 timer |
| C1 | 0.22μF capacitor |
| R1 | 10kΩ potentiometer |
| R2 | 1kΩ 1/4W resistor |
| R3 | 2.2kΩ 1/4W resistor |

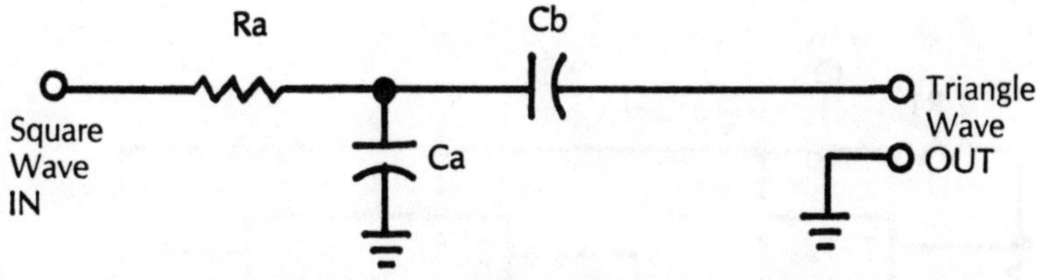


*Fig. 4-4* *This simple filter network can be used to modify the sound of the VCO.*

**Table 4-3 Parts List for the Filter Modification to Project 11—VCO (Voltage-Controlled Oscillator).**

| Part | Component Needed |
|---|---|
| Ca | 0.022μF capacitor |
| Cb | 0.001μF capacitor |
| Ra | 12kΩ 1/4W resistor |

For a little variety, add the simple filter circuit shown in Fig. 4-4 to the output of the VCO. This filter will weaken the harmonic content, making the output waveform resemble a triangular wave if the duty cycle is reasonably close to 1 to 2. You can achieve this effect by making resistor R2 small compared to resistor R1. A parts list for the filter modification is in Table 4-3.

## PROJECT 12: TWO-TONE OSCILLATOR

The circuit shown in Fig. 4-5 produces a two-tone signal (see Table 4-4 for the parts list). That is, the output will alternate between two frequencies.

There are two oscillators, or astable multivibrators, here. IC1 is free-running and is similar to project 10. This oscillator should have a low frequency (below the audible range); 0.5 to 5 Hz would be best.

The second oscillator (IC2) is a VCO like project 11. The control voltage input comes from the output of the first oscillator (IC1). This signal is LOW for a while, then it goes HIGH, and then it goes LOW again. This forces the VCO stage to alternate between two discrete frequencies.

For some real odd effects, experiment with a higher frequency control voltage. If the IC1 oscillator is operating in the audible range (about 20 Hz or so), you will create FM (frequency modulation) sidebands, resulting in a very complex and unusual

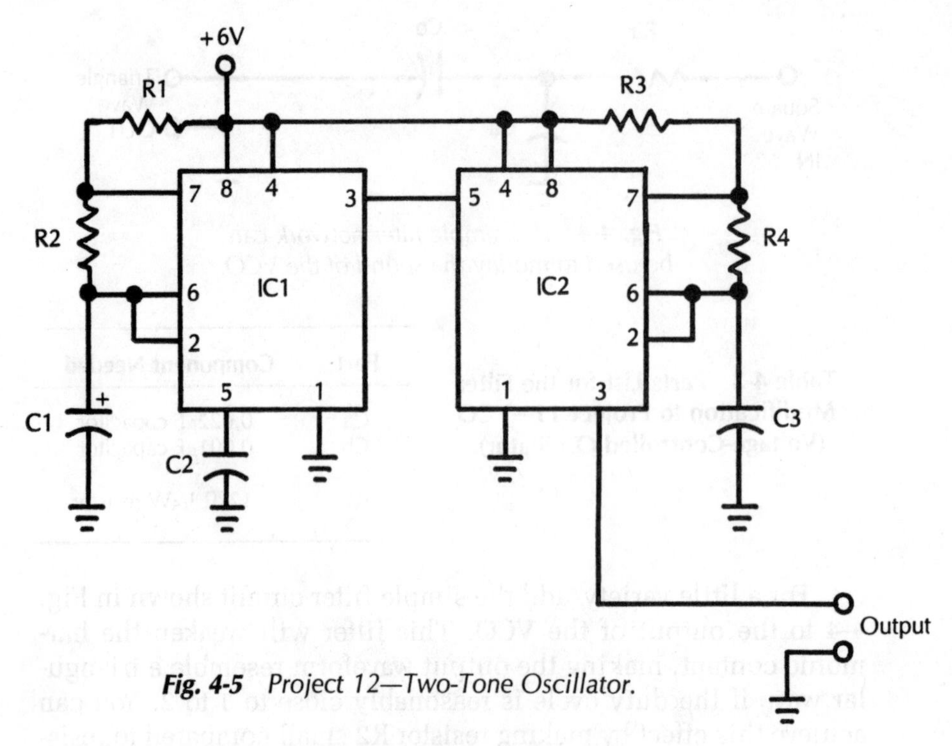


*Fig. 4-5* *Project 12—Two-Tone Oscillator.*

**Table 4-4 Parts List for Project 12—Two-Tone Oscillator.**

| Part | Component Needed |
|---|---|
| IC1, IC2 | 555 timer |
| C1 | 10$\mu$F 15V electrolytic capacitor* |
| C2 | 0.01$\mu$F capacitor |
| C3 | 0.22$\mu$F capacitor* |
| R1 | 220kΩ 1/4W resistor* |
| R2 | 1kΩ 1/4W resistor* |
| R3 | 12kΩ 1/4W resistor* |
| R4 | 2.2kΩ 1/4W resistor* |

*Experiment with other component values.

waveform at the output of IC2. If the control oscillator (IC1) has a higher frequency than the VCO (IC2), the effect is decidedly strange.

Whether the effect is musical or not is a judgment call left entirely to you. In any case, it is a lot of fun to experiment with this circuit. Try substituting different values for any or all of the components. Nothing is particularly critical in this circuit. However, the value of stabilizing capacitor C2 will not have a noticeable effect on the operation of this circuit.

## PROJECT 13: BIRD CHIRPER

The 3909 is a simple specialized IC. It contains the circuitry for a low-frequency oscillator. All that you need is an external capacitor to determine the actual frequency. The pin-out diagram for this device appears in Fig. 4-6. The intended application is a super-simple LED flasher. The 3909 is referred to as a *flasher/oscillator.*

It would be reasonable for you to wonder why the 3909 is discussed here, instead of in chapter 3. Well, as a LED flasher, the 3909 is a little too straightforward. It does not make a very interesting project, as the schematic diagram of Fig. 4-7 demonstrates.

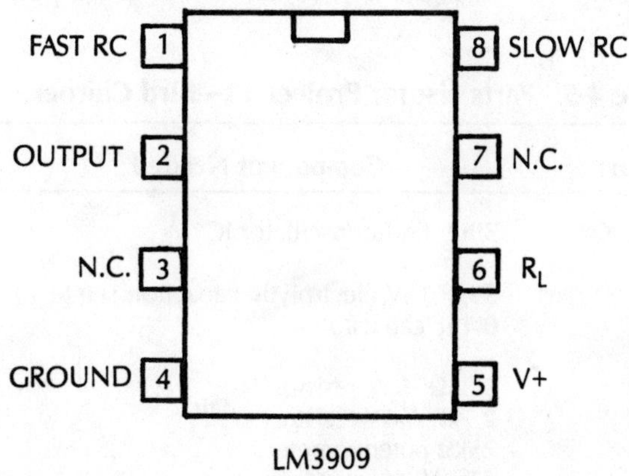


***Fig. 4-6*** *The 3909 is a dedicated flasher/oscillator integrated circuit.*

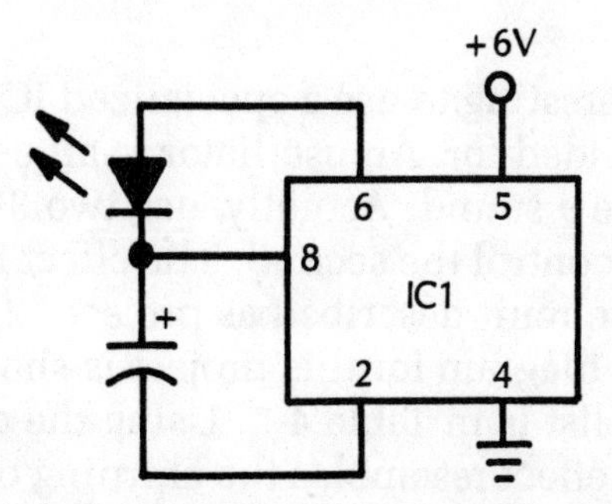


***Fig. 4-7*** *An LED flasher circuit built around the 3909 is very simple.*

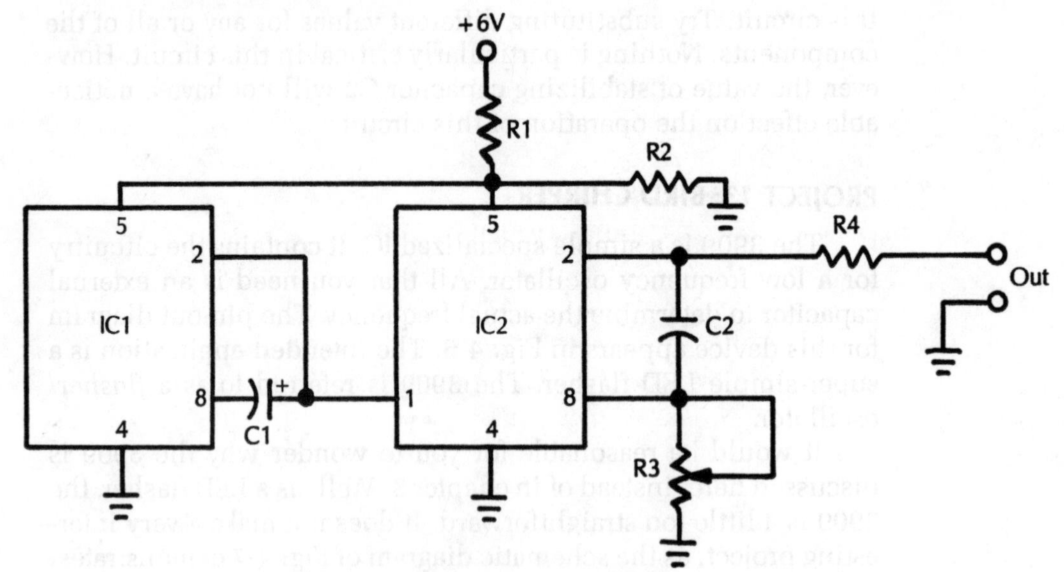


*Fig. 4-8* Project 13—Bird Chirper.

**Table 4-5 Parts List for Project 13—Bird Chirper.**

| Part | Component Needed |
|---|---|
| IC1, IC2 | 3909 flasher/oscillator IC |
| C1 | 33μF 15V electrolytic capacitor; see text. |
| C2 | 0.1μF capacitor |
| R1 | 3.3kΩ 1/4W resistor |
| R2 | 2.7kΩ 1/4W resistor |
| R3 | 25kΩ potentiometer |
| R4 | 47Ω 1/4W resistor |

Often it is more interesting to use a specialized IC for an application it was not intended for. An oscillator is an oscillator, so use the 3909 to produce a sound. Actually, use two 3909 oscillators. One oscillator will control the second. The effect is similar to the two-tone oscillator circuit described as project 12.

The schematic diagram for this project is shown in Fig. 4-8, and a typical parts list is in Table 4-5. Using the component values listed here, the effect resembles the chirping of a bird. Experiment with other component values, especially capacitor C1.

Oscillator A (IC1) turns oscillator B (IC2) on and off, producing the chirping effect. The values of capacitor C2 and resistor R3 determine the tone frequency. Capacitor C1 sets the chirp rate. The smaller the value of this capacitor, the faster the chirping. If C1 is small enough, oscillator A will start operating in the audible region. Frequency modulation effects will start to show up in the output signal. Instead of a chirping sound, you will get a continuous, complex tone.

## PROJECT 14: TOY ORGAN

With a little imagination, you can turn an oscillator into a musical instrument, albeit a very simple one. In this project, use the simple digital oscillator circuit shown in Fig. 4-9 as the heart of a toy organ. The frequency of this oscillator is determined by the values of resistor R2 and capacitor C1. By switching in and out various values for one or both of these components, you can create different notes. Use the resistor as the main frequency (note-determining) component, because it is more convenient to use variable resistances than variable capacitances. Select between two capacitors, however. This switching will determine the overall range of the instrument.

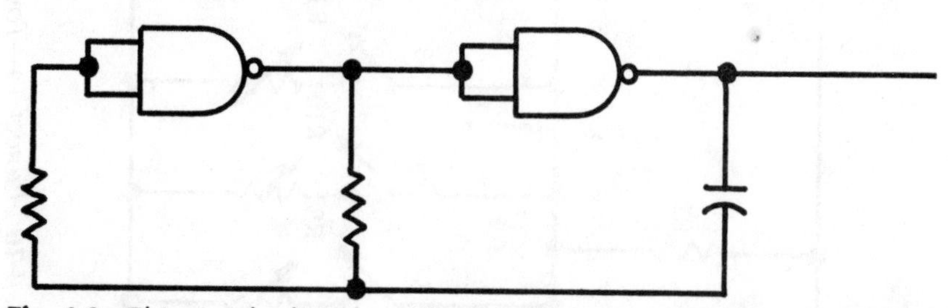

***Fig. 4-9*** *This simple digital oscillator is the heart of the toy organ project.*

The schematic diagram for this toy organ project appears in Fig. 4-10, and the parts list is in Table 4-6. This toy organ is a simple project, and a pretty crude musical instrument. But it is fun to experiment with, and it might make a nice gift for a child.

The frequency-determining resistances (R5 through R12) are switched in and out of the circuit using a keyboard made up of several simple normally open, *SPST* (single-pole, single-throw) push-button switches. The toy organ is strictly *monophonic*. That is, it can only sound a single note or tone at a time. It cannot

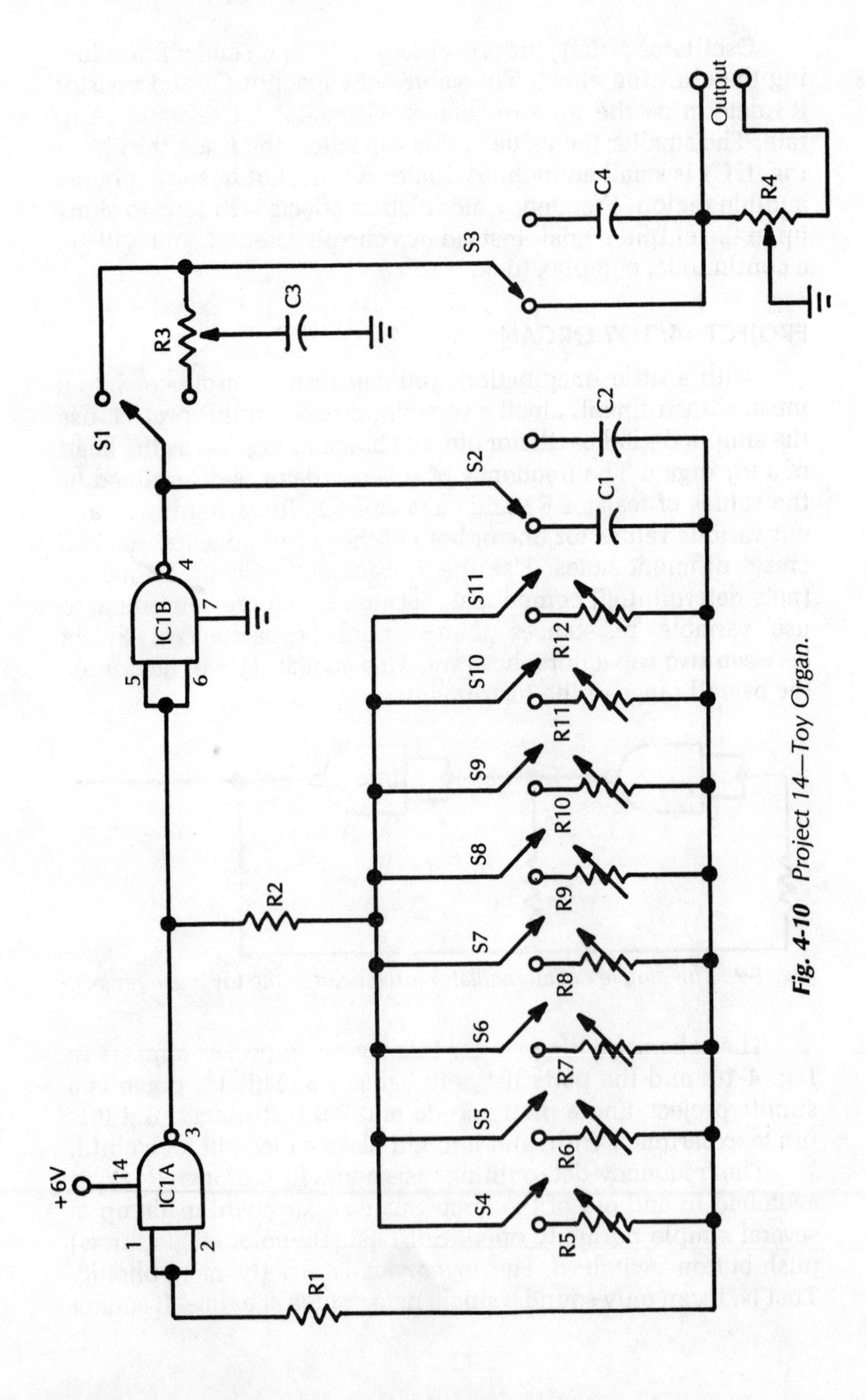


**Fig. 4-10** *Project 14—Toy Organ.*

**Table 4-6 Parts List for Project 14—Toy Organ.**

| Part | Component Needed |
|---|---|
| IC1 | CD4011 quad NAND gate |
| C1 | 0.01μF capacitor |
| C2 | 0.047μF capacitor |
| C3, C4 | 0.1μF capacitor |
| R1 | 1MΩ 1/4W resistor |
| R2 | 10kΩ 1/4W resistor |
| R3, R4 | 500kΩ potentiometer |
| R5–R12 | 250kΩ trimmer potentiometer* |
| S1, S2, S3 | *SPDT* (single-pole, double-throw) switch |
| S4–S12 | Normally open SPST push switch* |

*Add more for additional notes, see text.

play chords. If you close two or more key switches simultaneously, the frequency will be thrown off. All of the selected resistors are in parallel, lowering their overall effective value.

Use trimmer potentiometers (R5 through R12) for each individual note switch to permit you to tune the notes. If you do not care about standard tuning, substitute fixed resistors for the trimmer potentiometers.

Only 8 keys (notes) are shown in the schematic, but you can easily expand the keyboard to cover as many notes as you like. Just add more key switches and trimmer potentiometers.

If no key switches (S4 through S11) are closed, the frequency-determining resistance will effectively be infinite (or at least, extremely large), so the oscillator output frequency will be well outside the audible range. This is a pretty crude approach, but it is cheap and works fairly well.

Use switch S2 to select between two frequency-determining capacitors (C1 and C2). Switch S2 functions as a range switch.

Switch S1 selects or bypasses a simple passive filter made up of resistor R3 and capacitor C3. Similarly, capacitor C4 is a simple filtering device that you can switch in or out of the circuit via switch S3. Potentiometer R4 serves as a volume control for the toy organ.

## PROJECT 15: FREQUENCY DIVIDER

Suppose you have a signal that has too high a frequency for some particular application. This often happens in large systems, where a high frequency is required by some stages, while other stages may have a slower response. Frequencies might also need to be lowered in an electronic music system. In any case, the solution is to use a frequency divider.

Figure 4-11 shows how the 555 timer IC can be used as a straightforward frequency divider, and a typical parts list for this project appears as Table 4-7.

Essentially, this circuit is just a basic monostable multivibrator circuit. For reliable operation, the input signal should be a rectangular (or square) wave, or a string of pulses. One of the input pulses triggers the timer. Until it times out, the circuit puts out a HIGH level output. Additional input pulses are ignored during the timing cycle. Once the timing period cycles out, the 555 output will go LOW until the next input pulse is received. This frequency divider circuit will alter the duty cycle of the waveform.

Experiment with other values of resistor R1 and capacitor C1. The values of these components set up the timing cycle period. The length of the timing cycle will control how much the input frequency is reduced. For proper operation, the timing cycle period must be longer than the input pulses.

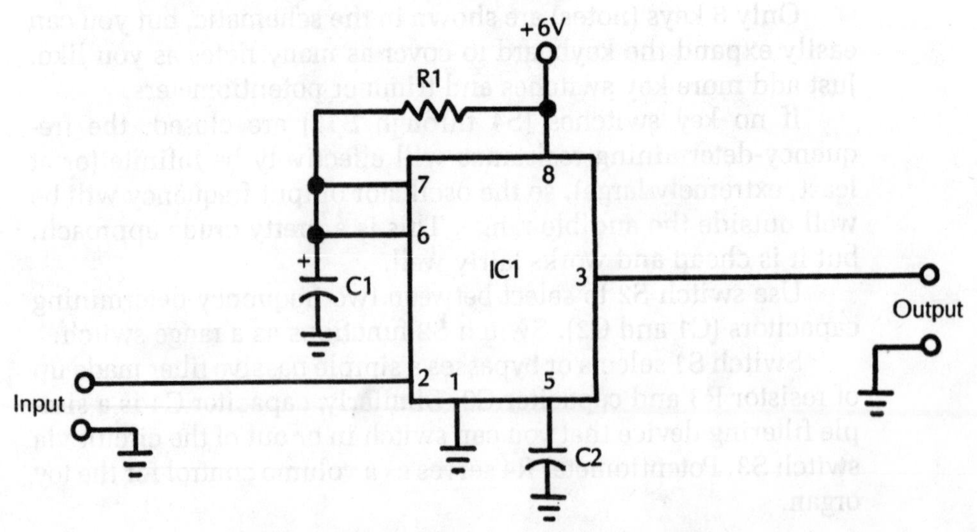


*Fig. 4-11* *Project 15—Frequency Divider.*

**Table 4-7 Parts List for Project 15—Frequency Divider.**

| Part | Component Needed |
|---|---|
| IC1 | 555 (or 7555) timer |
| C1 | 1μF 15V electrolytic capacitor* |
| C2 | 0.01μF capacitor |
| R1 | 470kΩ 1/4W resistor* |

*See text.

## PROJECT 16: FREQUENCY MULTIPLIER

This project works just the opposite of project 15. This is a frequency multiplier. It increases the input frequency. The circuit is shown in Fig. 4-12, and the parts list, which is quite simple, is in Table 4-8.

There are two outputs to this frequency multiplier circuit. If the input frequency is F, the frequency at output A will be 2F, and the frequency at output B will be 4F. Because digital gates are used in this circuit, the input signal must be a rectangular or pulse wave. Other waveforms might or might not work reliably with this circuit.

There is not much of anything to experiment with other than component values within this project. Changing the values of either of two resistors will not do anything interesting, but if these resistances are made too large or too small, the project might not work properly.

## PROJECT 17: METRONOME

A metronome is useful for a musician who is rehearsing. It helps keep a steady beat. A useful, but simple, metronome circuit is shown in Fig. 4-13, and the parts list is in Table 4-9. This project is really quite simple. It is basically two oscillators.

IC1, a 555 timer, is the main, time-keeping oscillator. It is set for a low frequency. The duty cycle is also set for a very narrow pulse. Potentiometer R1 allows you to manually adjust the pulse frequency. A large knob with a calibrated scale would be very helpful. The scale should be calibrated in (beats per minute).

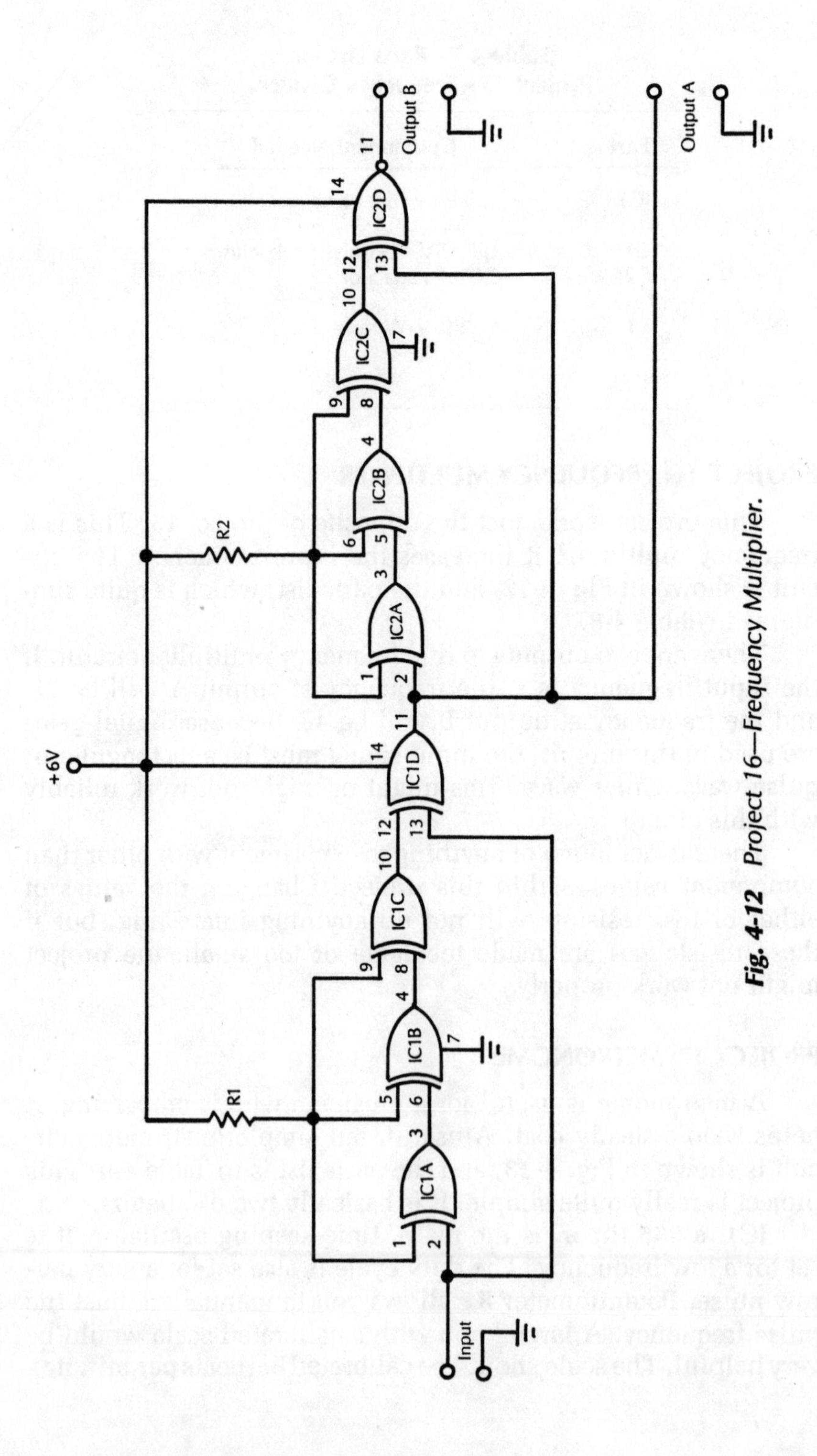


**Fig. 4-12** *Project 16—Frequency Multiplier.*

**Table 4-8 Parts List for Project 16—Frequency Multiplier.**

| Part | Component Needed |
|---|---|
| IC1, IC2 | CD4070 quad XOR gate |
| R1, R2 | 2.2kΩ 1/4W resistor |

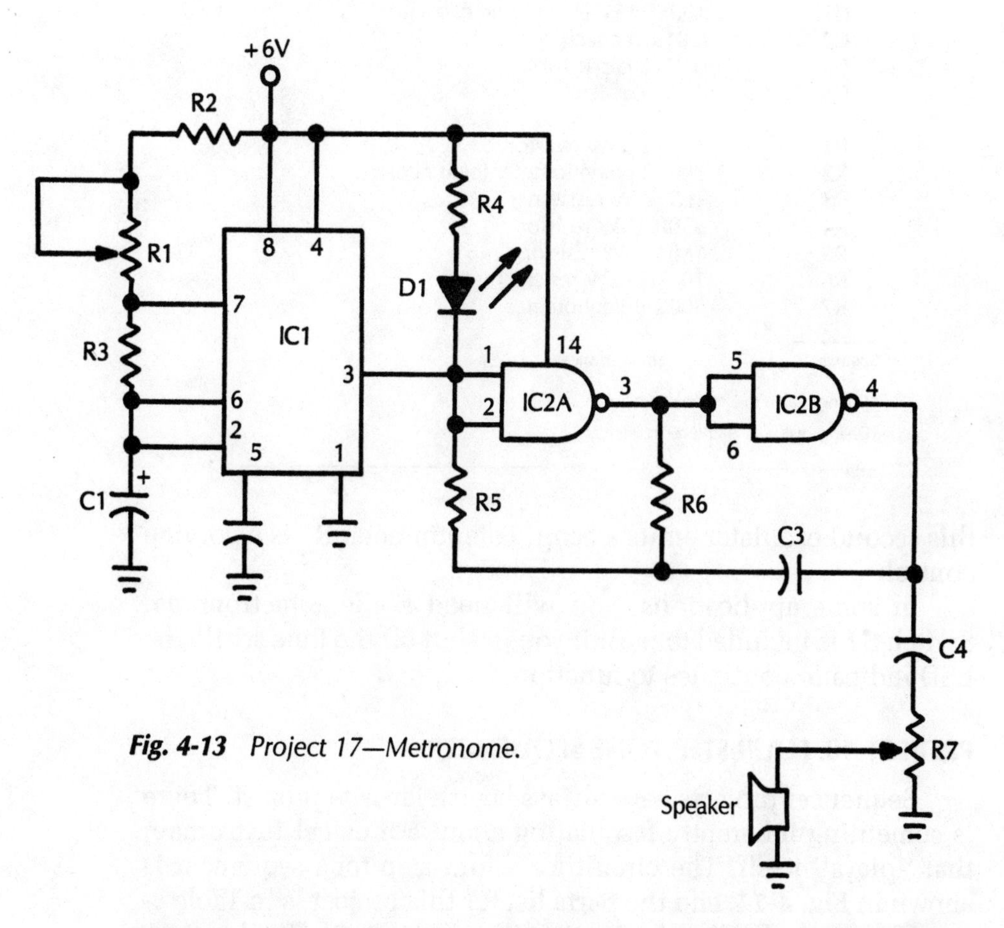


*Fig. 4-13 Project 17—Metronome.*

It can be calibrated with a stop watch, or a dial-face watch with a sweep second hand.

On each pulse, the LED flashes on briefly. The timing pulses also gate the second oscillator. This oscillator, built around IC2, is a gated audio frequency oscillator. It is similar (except for the frequency) to the circuit used back in project 6. On each pulse,

**Table 4-9 Parts List for Project 17—Metronome.**

| Part | Component Needed |
|---|---|
| IC1 | 555 (or 7555) timer |
| IC2 | CD4011 quad NAND gate |
| D1 | LED |
| C1 | 100μF 15V electrolytic capacitor* |
| C2 | 0.01μF capacitor |
| C3 | 0.01μF capacitor† |
| C4 | 0.1μF capacitor |
| R1 | 470kΩ 1/4W resistor |
| R2 | 500kΩ potentiometer (beat rate) |
| R3 | 1kΩ 1/4W resistor |
| R4 | 330Ω 1/4W resistor |
| R5 | 1MΩ 1/4W resistor |
| R6 | 100kΩ 1/4W resistor† |
| R7 | 500Ω potentiometer |
| Speaker | Small speaker |

*Change value to alter beat rate range.

†Change value to alter tone frequency.

this second oscillator emits a beep. Potentiometer R7 is a volume control.

In some applications, you will need a silent metronome. Switch S1 is included to permit you to shut off the tone while the LED indicator continues to function.

## PROJECT 18: FOUR-STEP TONE SEQUENCER

Sequencer projects have always been a favorite project. There is something inherently fascinating about a musical instrument that "plays" itself. The circuit for a four-step tone sequencer is shown in Fig. 4-14, and the parts list for this project is in Table 4-10. When completed, this circuit will play a user-defined pattern of four notes over and over, automatically.

Use four gated oscillators to provide the tones. A potentiometer, or trimmer potentiometer, permits you to tune each oscillator

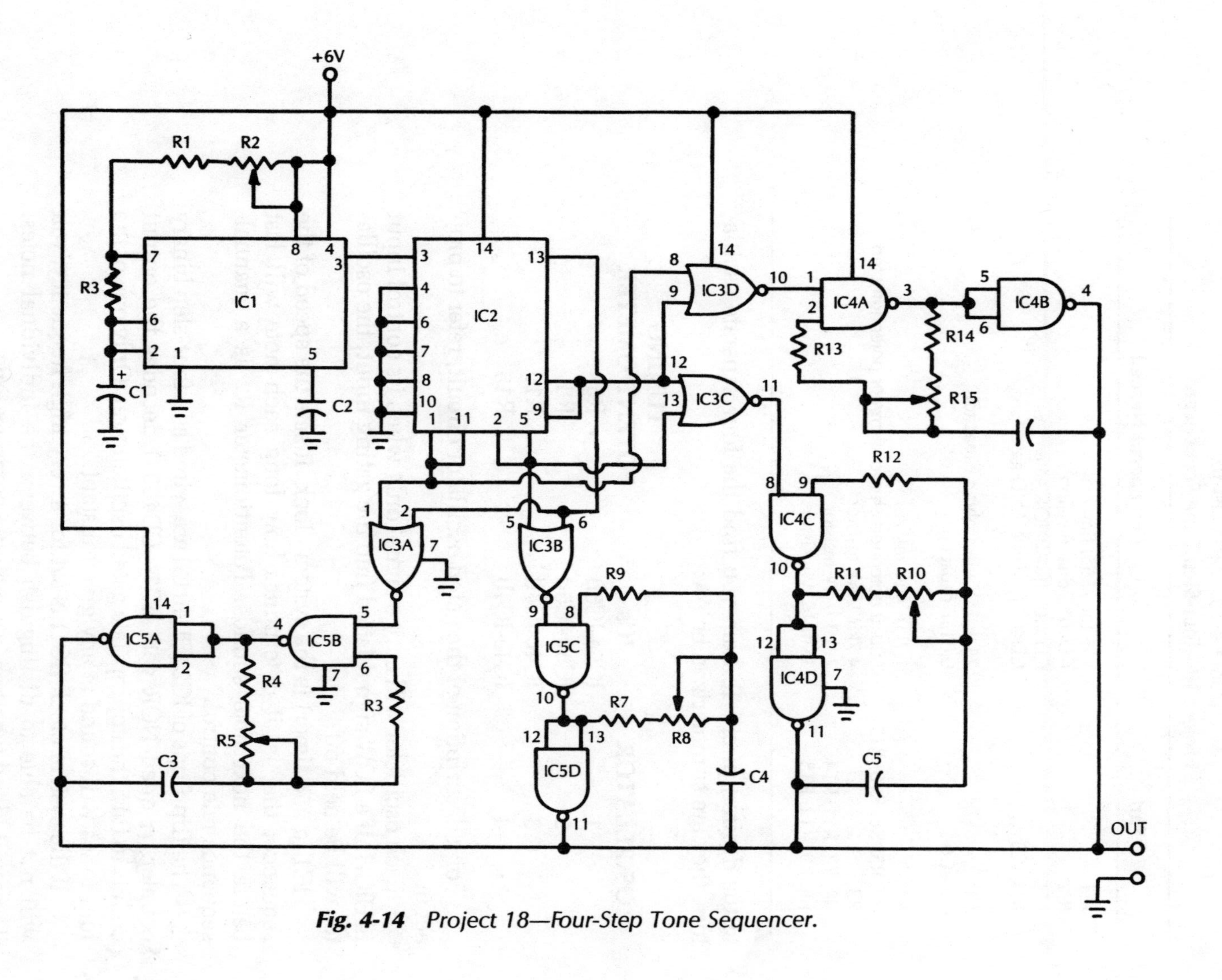


**Fig. 4-14** *Project 18—Four-Step Tone Sequencer.*

**Table 4-10 Parts List for Project 18—Four-Step Tone Sequencer.**

| Part | Component Needed |
|---|---|
| IC1 | 555 (or 7555) timer |
| IC2 | CD4013 dual flip-flop |
| IC3 | CD4001 quad NOR gate |
| IC4, IC5 | CD4011 quad NAND gate |
| C1 | 25μF 15V electrolytic capacitor |
| C2–C6 | 0.01μF capacitor |
| R1 | 1.2kΩ 1/4W resistor |
| R2, R5, R8, R10, R15 | 500kΩ potentiometer (or trimmer potentiometer) |
| R3 | 4.7kΩ 1/4W resistor |
| R4, R7, R11, R14 | 22kΩ 1/4W resistor |
| R6, R9, R12, R13 | 1MΩ 1/4W resistor |

to the desired pitch. To help you find the four separate oscillators, they are summarized below:

| OSCILLATOR | ICs | TUNING POTENTIOMETER |
|---|---|---|
| 1 | IC5a-IC5b | R5 |
| 2 | IC5c-IC5d | R8 |
| 3 | IC4c-IC4d | R10 |
| 4 | IC4a-IC4b | R15 |

For a description of the gated oscillator circuit, refer to project 6.

Each oscillator will emit a signal only when its control input is HIGH. If a LOW signal is fed into the gating input, the oscillator will be held off.

IC1 (a 555 timer) is the system clock. It sets the speed of the sequence; that is, it determines how long each note will last before the next note sounds. Potentiometer R2 is a manual-sequence rate control.

The flip-flops of IC2 (CD4013) are wired as a four-step binary counter. A quad NOR gate (IC3—CD4001) decodes the output counts to turn the appropriate gated oscillator on. Only one of the four tones will sound at any given instant.

If the system clock (IC1) is set for a very high frequency, you will not be able to distinguish between the individual notes. They will blend together into a single complex tone.

Nothing is critical about this circuit. Experiment with other component values for any or all of the passive components (resistors and capacitors) in the circuit. For the most interesting results, experiment with the values of capacitors C1 and C3 through C6. However, the timer stabilizing capacitor (C2) is not worth experimenting with. Changing its value will have no noticeable effect on circuit operation. In many cases it can be eliminated from the circuit altogether. It is cheap insurance against possible stability problems, so it is best to include it.

## PROJECT 19: COMPLEX TONE GENERATOR

If you like creating unusual sounds, you should enjoy the circuit of Fig. 4-15. The parts list is in Table 4-11. This is not a synthesizer. You probably will not be able to come close to simulating any natural sounds, but you can generate a wide variety of complex tones.

There are two tone generators in this circuit: IC1/IC2 and IC3. IC1 (555 timer) is an astable multivibrator or oscillator. Its frequency and duty cycle are controlled by potentiometers R2 and R3 and switch S3, which selects one of four timing capacitors (C1 through C4).

IC2 (555 timer) is a frequency divider, like the one in project 15. Its time constant can be adjusted with potentiometer R4. As discussed in the text for project 15, in addition to reducing the frequency, this circuit also changes the duty cycle. A number of unique sounds can be generated with just IC1 and IC2.

IC3 (555 timer) is another astable multivibrator or oscillator. It is quite straightforward. The frequency of this oscillator is controlled with potentiometer R7. Remember, changing the frequency also changes the duty cycle.

Finally, there is IC4 (CD4011 quad NAND gate). These gates combine the outputs of IC2 and IC3. IC4A is a NAND gate. IC4b is wired as an inverter. Together, they function as an AND gate. The output is HIGH if, and only if, both inputs are HIGH. For most settings of the controls, the final output will not sound like a musical tone. It will be a raspy, very odd sound of some sort. There are countless variations in this sound, depending on the settings of the various controls.

Experiment with the values of capacitors C6 and C8. If you use a rotary switch with more than four positions for S1, you can add more capacitor values along with C1 through C4.

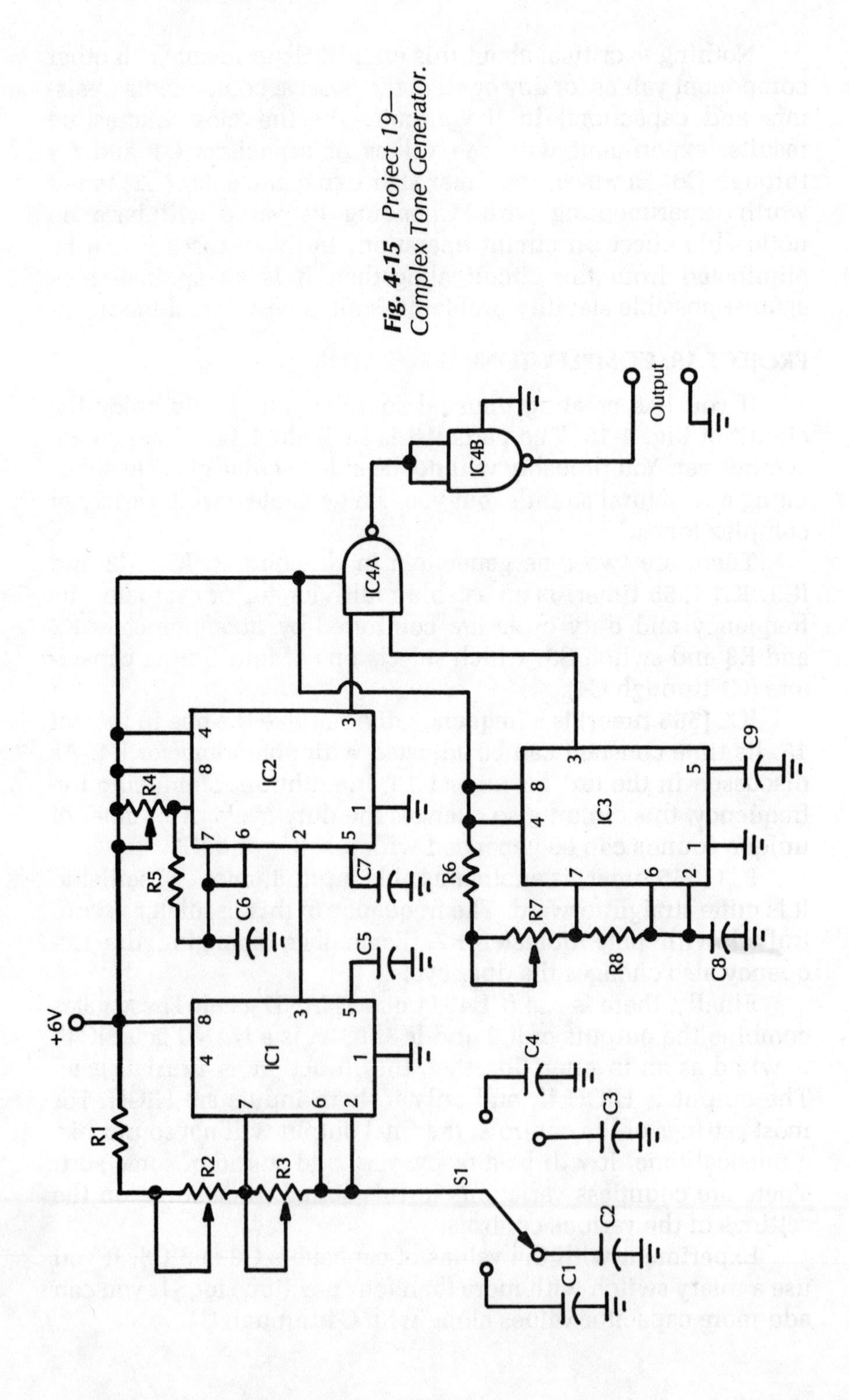


***Fig. 4-15*** *Project 19—Complex Tone Generator.*

**Table 4-11 Parts List for Project 19—Complex Tone Generator.**

| Part | Component Needed |
|---|---|
| IC1, IC2, IC3 | 555 (or 7555) timer |
| IC4 | CD4011 quad NAND gate |
| C1, C5, C7, C9 | 0.01μF capacitor |
| C2 | 0.05μF capacitor |
| C3 | 0.1μF capacitor |
| C4 | 0.5μF capacitor |
| C6 | 0.022μF capacitor |
| C8 | 0.1μF capacitor |
| S1 | SP4T rotary switch; see text. |

Do not bother experimenting with capacitors C5, C7, and C9. These capacitors are used for stability insurance only. Their values do not have a direct effect on circuit operation.

Have fun with this circuit. Just feed the output through any audio amplifier and speaker. But keep the volume down. Neighbors and other family members might be less than enthralled with the results of your experimentation.

# 5 Switching and Timing Circuits

THE PROJECTS IN CHAPTERS 3 AND 4 ARE FUN. YOU BUILD THEM, AND they do something direct and interesting. Other electronic circuits are somewhat more mundane, at least when used by themselves. They can, however, be combined into sophisticated and exciting systems of all types.

One of the most common electronic functions is switching. The projects in this chapter all perform some switching or timing operation. Combine these projects with other circuits, and you will really have something exciting.

## PROJECT 20: TOUCH SWITCH

Touch switches are popular projects. Any electrical or electronic device can be operated by nothing more than a light touch of a fingertip. Usually, the touch switch itself is nothing more than a sensitive detection circuit wired to a pair of conductive plates. Two small pieces of copper-clad (nonetched PC) board will do nicely. There is a very small space between the two touch plates. The plates are positioned so that they can be bridged easily by your fingertip. The operator's body resistance through the finger effectively provides a current path between the plates. This is the equivalent of closing the contacts of a mechanical switch.

**IMPORTANT!!!**

**Because the human body serves as part of the current path, operate all touch switches by battery power ONLY. Never use an ac power supply with any touch-switch project.**

You might think it would be perfectly safe to use a well-regulated ac-to-dc power supply, like those presented in chapter 1 of this book. After all, there will only be a low-power dc voltage fed into the touch-switch circuit. Right? Well, normally the answer would be yes. But things are not always normal. There is always a chance for an unexpected short circuit that allows a high-voltage ac signal to reach the touch plates. The result could be a painful electrical shock or even death. Please, believe it. It is not worth the risk. Use battery power only!

Similarly, if the device to be controlled by the touch-switch circuit is ac powered, the controlled device must be 100 percent isolated from the switching circuit. Use a relay, or an opto-isolator so that there is no possible path for any ac power signal to get back into the touch plates under any conditions, no matter how unlikely.

A simple, practical touch-switch circuit is shown in Fig. 5-1, and the parts list for this easy project is in Table 5-1. Switch S1 is an enable switch. You can omit it if you prefer. When this switch is open (or missing), pin #1 of IC1 is grounded through resistor R3. This forces a logic LOW input at this pin. Because IC1 is a NAND gate, a LOW signal on pin #1 will cause the output to always be HIGH, regardless of the input state of pin #2. This

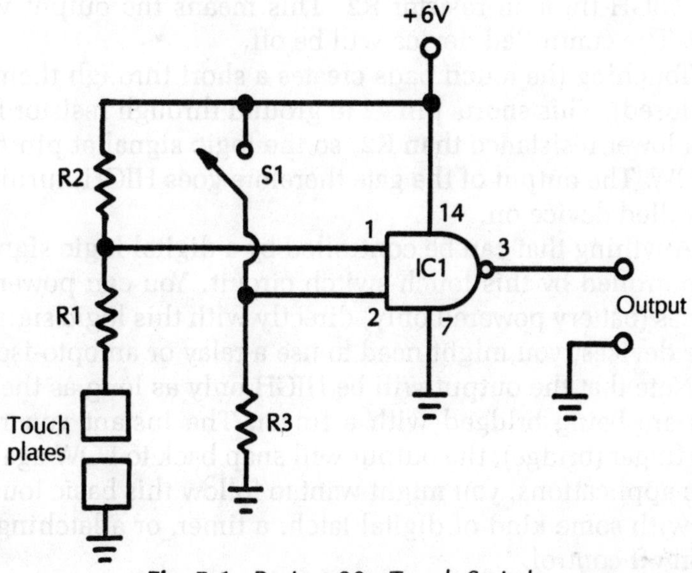


***Fig. 5-1*** *Project 20—Touch Switch.*

**Table 5-1 Parts List for Project 20—Touch Switch.**

| Part | Component Needed |
|---|---|
| IC1 | CD4011 quad NAND gate |
| R1 | 100kΩ 1/4W resistor |
| R2, R3 | 10MΩ 1/4W resistor |

allows you to bypass the touch switch and operate the controlled device normally.

Closing switch S1 enables the touch-switch circuit. The logic signal at pin #1 will be HIGH. The logic state at pin #2 will control the output at pin #3. The signal will be inverted by the gate. That is, if pin #2 is HIGH, the output will be LOW; but, if pin #2 is LOW, then the output will be HIGH.

If you decide to omit the enable switch (S1), you must also delete resistor R3 from the circuit, and connect pin #1 of IC1 directly to the positive supply voltage line. If you do not make these modifications, the project will not work.

Assume the touch switch is enabled. Either switch S1 is closed or S1 and R3 have been removed from the circuit. When the touch pads are not being touched (they are open), pin #2 is held HIGH through resistor R2. This means the output will be LOW. The controlled device will be off.

Touching the touch pads creates a short through them (they are closed). This shorts pin #2 to ground through resistor R1. R1 has a lower resistance than R2, so the logic signal at pin #2 will be LOW. The output of the gate therefore goes HIGH, turning the controlled device on.

Anything that can be controlled by a digital logic signal can be controlled by this touch-switch circuit. You can power some devices (battery powered only) directly with this logic signal. For other devices, you might need to use a relay or an opto-isolator.

Note that the output will be HIGH only as long as the touch pads are being bridged with a finger. The instant you remove your finger (bridge), the output will snap back to LOW again. For some applications, you might want to follow this basic touch circuit with some kind of digital latch, a timer, or a latching relay, for on/off control.

## PROJECT 21: TIMED TOUCH SWITCH

The touch switch of project 20 responds instantly. As soon as you bridge the touch plates with your finger, you turn the controlled device on. As soon as you remove your finger from the touch plates, the controlled device is turned off. The circuit shown in Fig. 5-2 adds a short delay to the switching operation. The parts list for this project appears in Table 5-2. This touch switch circuit features a built-in monostable multivibrator or timer.

The output of this circuit is normally LOW. Bridging a fingertip across the touchplates triggers the timer. The output goes HIGH for a predetermined period. The length of time the touch plates are bridged by the finger is irrelevant.

For proper operation, use identical values for resistors R1 and R3. Call that value *R*. The delay time is determined by the value of *R*, and the value of capacitor C1. Find approximate time period simply by multiplying these two values:

$$T = RC$$

For most reliable operation, 100kΩ is a good value to use for R. This is the value called for in the parts list. Change the value of capacitor C1 to alter the length of the output pulse.

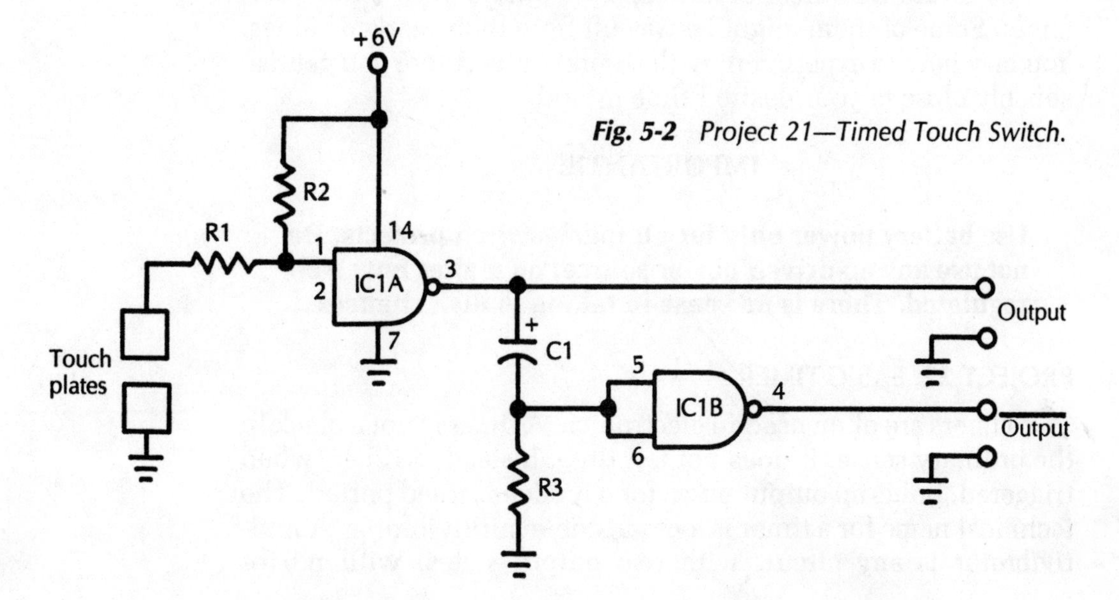


*Fig. 5-2 Project 21—Timed Touch Switch.*

**Table 5-2 Parts List for Project 21—Timed Touch Switch.**

| Part | Component Needed |
|---|---|
| IC1 | CD4011 quad NAND gate |
| C1 | 22μF 15V electrolytic capacitor; see text. |
| R1, R3 | 100kΩ 1/4W resistor |
| R2 | 10MΩ 1/4W resistor |

Assuming that R is 100kΩ, estimate the desired value of C1 just by multiplying the desired time (in seconds) by 10. This will give you the approximate capacitance needed in microfarads (μF):

$$C1 = 10T$$

Using the 22μF capacitor called for in the parts list will result in an output pulse of a little over two seconds. The time period is limited. Practical capacitance values will keep it low. For an output pulse of about 1 minute (60 seconds), try combining a 470μF capacitor and a 150μF capacitor in parallel.

Be aware that electrolytic capacitors have very wide tolerances. Some of them might be way off from their marked values. You may have to experiment with several units before you get reasonably close to your desired time period.

**IMPORTANT!!!**

**Use battery power only for all touch-switch projects. Do not use any ac-driven power source, no matter how well regulated. There is no sense in taking foolish chances.**

## PROJECT 22: BASIC TIMER

Timers are often used in electronics. A timer is not a clock in the ordinary sense. It does not tell time. Instead, a timer, when triggered, holds an output pulse for a predetermined period. The technical name for a timer is a *monostable multivibrator*. A *multivibrator* is any circuit with two output states, with no (or extremely little) transition between them. The two output states are most commonly labelled HIGH and LOW.

A monostable multivibrator has one stable output state and one unstable output state. Assume that the stable output state is LOW. When there is no trigger pulse at the input, nothing will happen. The output will remain in its stable state (LOW). When a trigger pulse is detected at the input, however, the output will jump to the opposite, unstable state (HIGH). This state will be held for a fixed period of time, dependent on specific resistances and capacitances within the timer circuit. After the preset time has expired, the output signal will revert to its original, stable state (LOW). (Some monostable multivibrators use HIGH as the normal, stable output state, and go LOW during the timing period.)

The timing period is fixed by component values. It is not influenced by the length of the input trigger pulse. Because the output pulse is normally longer than the input trigger pulse, this type of circuit is sometimes referred to as a *pulse stretcher*.

The 555 timer is one of the most popular ICs around. It is designed specifically for monostable and astable multivibrator circuits. A CMOS version of the 555 is also available. The CMOS version is called the 7555.

The basic 555 timer circuit is shown in Fig. 5-3, and the parts list is in Table 5-3. The timer is triggered by a negative going pulse at pin #2. The 555 senses when this input signal switches from HIGH to LOW. An optional RESET input is also provided at pin #4. An input pulse at this pin will abort the current timing sequence. The output will revert to LOW, even if the timing cycle has not been completed.

Once the timing cycle is over (or the timer has been reset), the 555 is ready to detect another input pulse at the TRIGGER input. The time period is determined by the values of resistor R1 and capacitor C1. The equation is:

$$1.1(R1)(C1)$$

where R1 is measured in ohms, C1 is measured in microfarads, and T is the time in seconds. For the component values in Table 5-3, the timing cycle will last approximately five seconds. Experiment with other component values.

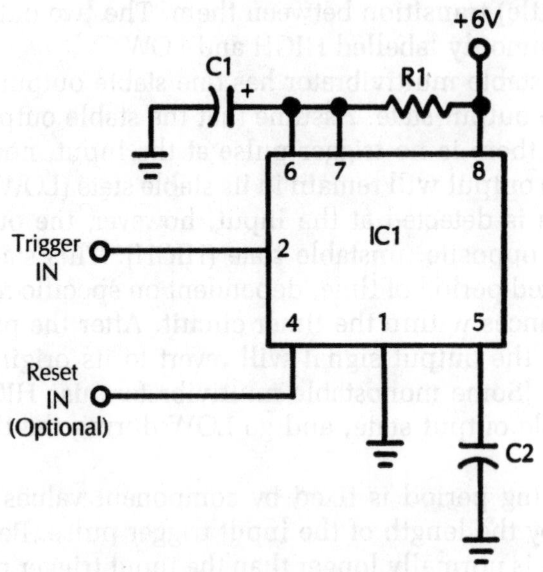


*Fig. 5-3* Project 22—Basic Timer.

**Table 5-3 Parts List for Project 22—Basic Timer.**

| Part | Component Needed |
|---|---|
| IC1 | 555 (or 7555) timer |
| C1 | 22μF 15V electrolytic capacitor* |
| C2 | 0.01μF capacitor |
| R1 | 220kΩ 1/4W resistor* |

*Timing components; see text.

Capacitor C2 is included to stabilize the 555 timer chip. This capacitor will not be necessary in all cases, but it is cheap insurance against possible stability problems. Include this capacitor in all of your circuits using the 555 (unless, of course, pin #5 is used for other purposes). The exact value of this capacitor is not at all critical. Changing the value of C2 will not directly affect circuit operation in any way. There is no point in experimenting with other values for this particular component.

The circuit can be triggered by any digital or pulse signal source. Figure 5-4 shows a simple circuit for manually triggering

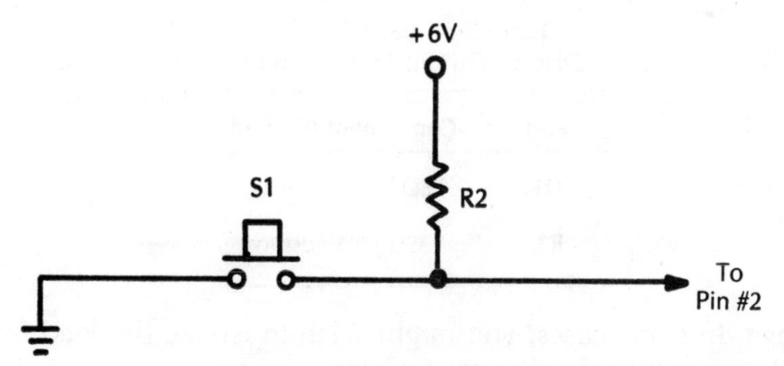


*Fig. 5-4* The basic timer circuit of Fig. 5-3 can be manually triggered with this network.

**Table 5-4 Parts List for Manual Trigger Modification.**

| Part | Component Needed |
|---|---|
| R2 | 1MΩ 1/4W resistor |
| S1 | Normally open SPST push-button switch |

the timer, the parts list for this modification is in Table 5-4. This input network can also be used to control the RESET input (pin # 4).

Figure 5-5 shows how the output state can be indicated by an LED. The parts list for this modification is in Table 5-5. The LED will be dark when the timer input is LOW. When the timer input goes HIGH, the LED will light up until the timing cycle is over.

Use the output to drive any of many low-power analog and digital circuits. Anything that can be operated by a pulse signal can be controlled by the timer. For some loads, a buffer amplifier stage might be needed to supply sufficient current or to boost the

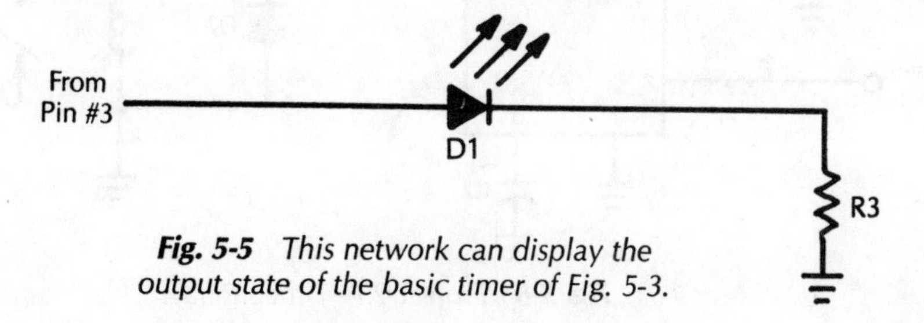


*Fig. 5-5* This network can display the output state of the basic timer of Fig. 5-3.

Table 5-5 Parts List for Display-Output Modification.

| Part | Component Needed |
|---|---|
| D1 | LED |
| R3 | 330Ω 1/4W resistor |

voltage. In some cases, you might wish to isolate the load from the timer with a relay or opto-isolator.

## PROJECT 23: TIMED RELAY

Almost any electrically powered device can be turned on and off by a timer such as is shown in Fig. 5-6. This project is essentially a variation of project 22. A typical parts list for this project is given in Table 5-6.

Using the component values shown here, the timing period will be approximately 15 seconds. Experiment with other values for resistor R1 and capacitor C1.

Diode D1 prevents any possible ac signal from reaching the relay (K1) coil. Diode D2 blocks any back-*EMF* (electromotive force), which could cause the coil to burn itself out. Almost any silicon diodes can be used for D1 and D2. The 1N4148 is listed just because it is probably the most widely available diode type.

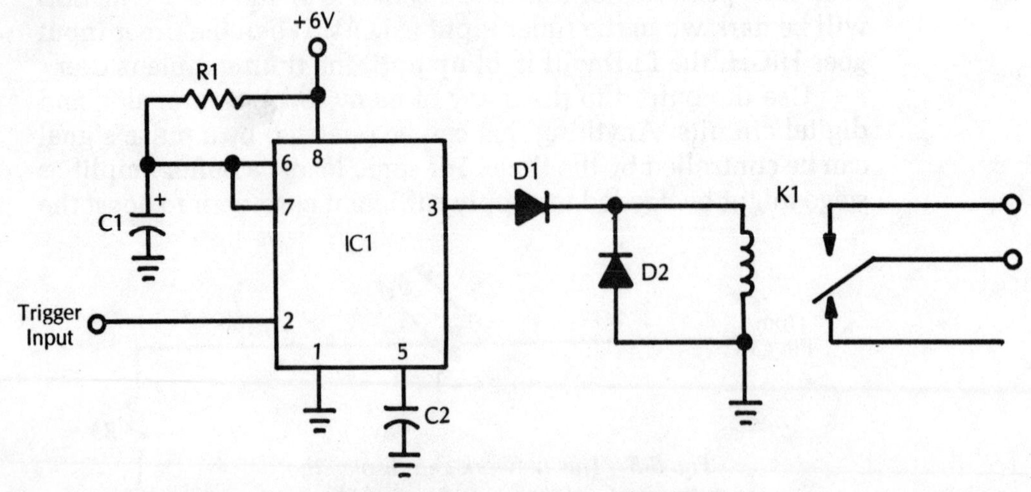


*Fig. 5-6* *Project 23—Timed Relay.*

**Table 5-6 Parts List for Project 23—Timed Relay.**

| Part | Component Needed |
|---|---|
| IC1 | 555 (or 7555) timer |
| D1, D2 | 1N4148 diode (or similar) |
| K1 | Relay to suit load |
| C1 | 50μF 15V electrolytic capacitor* |
| C2 | 0.01μF capacitor |
| R1 | 270kΩ 1/4W resistor* |

*Timing components; see text.

This component is sometimes numbered 1N914. The 1N914 and 1N4148 diodes are identical and can be interchanged in any circuit.

Of course, the load device is connected to the appropriate relay switching contacts. If the normally-open (N.O.) contacts are used, the output device will be on only during the timing cycle, and off at all other times. By using the normally-closed (N.C.) contacts, the load device can normally be on, and cut off only during the timing period.

## PROJECT 24: LONG-DURATION TIMER

The 558 is a quad timer. It contains four independent timer sections. The pin-out diagram for this chip appears in Fig. 5-7. Each of the internal timer sections is the equivalent to a somewhat stripped-down 555 timer. To minimize the number of pins, some 555 functions are eliminated. The 558 is designed solely for use in monostable multivibrator circuits. It is not intended for astable applications.

Each of the four timer stages is independent. They share only the power supply connections. The timer stages can be used separately, or together.

By stringing multiple timer stages together in series, one after another, as shown in Fig. 5-8, you can expand significantly the maximum timing period. A single 555 can only be used

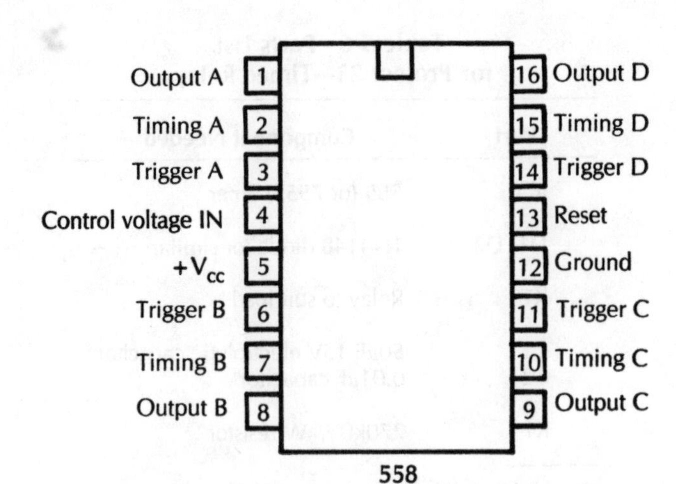


***Fig. 5-7*** *The 558 contains four independent timer stages in a single package.*

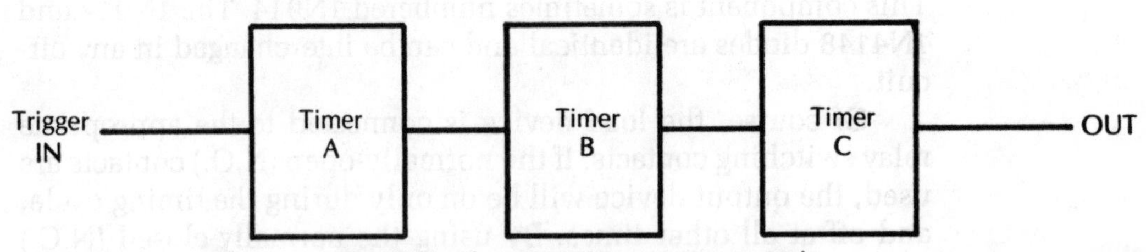


***Fig. 5-8*** *The maximum timing period can be extended by cascading timer stages.*

directly for time periods up to a few minutes. By using multiple timer sections, you can add the individual timing periods together.

The output of each stage is normally LOW. When the first stage is triggered by a negative-going pulse, its output goes HIGH for time period A. At the end of time period A, the first stage output goes LOW, placing a negative-going pulse on the TRIGGER input of the second stage. This initiates the second stage timing period. This process continues on down through the line.

A circuit for using a 558 quad timer in a four-stage, long-duration timer is shown in Fig. 5-9, and a typical parts list is given in Table 5-7. Using these component values, the time period for each individual stage is about 25.85 seconds. The total timing period of all four stages is approximately 103.4 seconds (or 1 minute and 43.4 seconds).

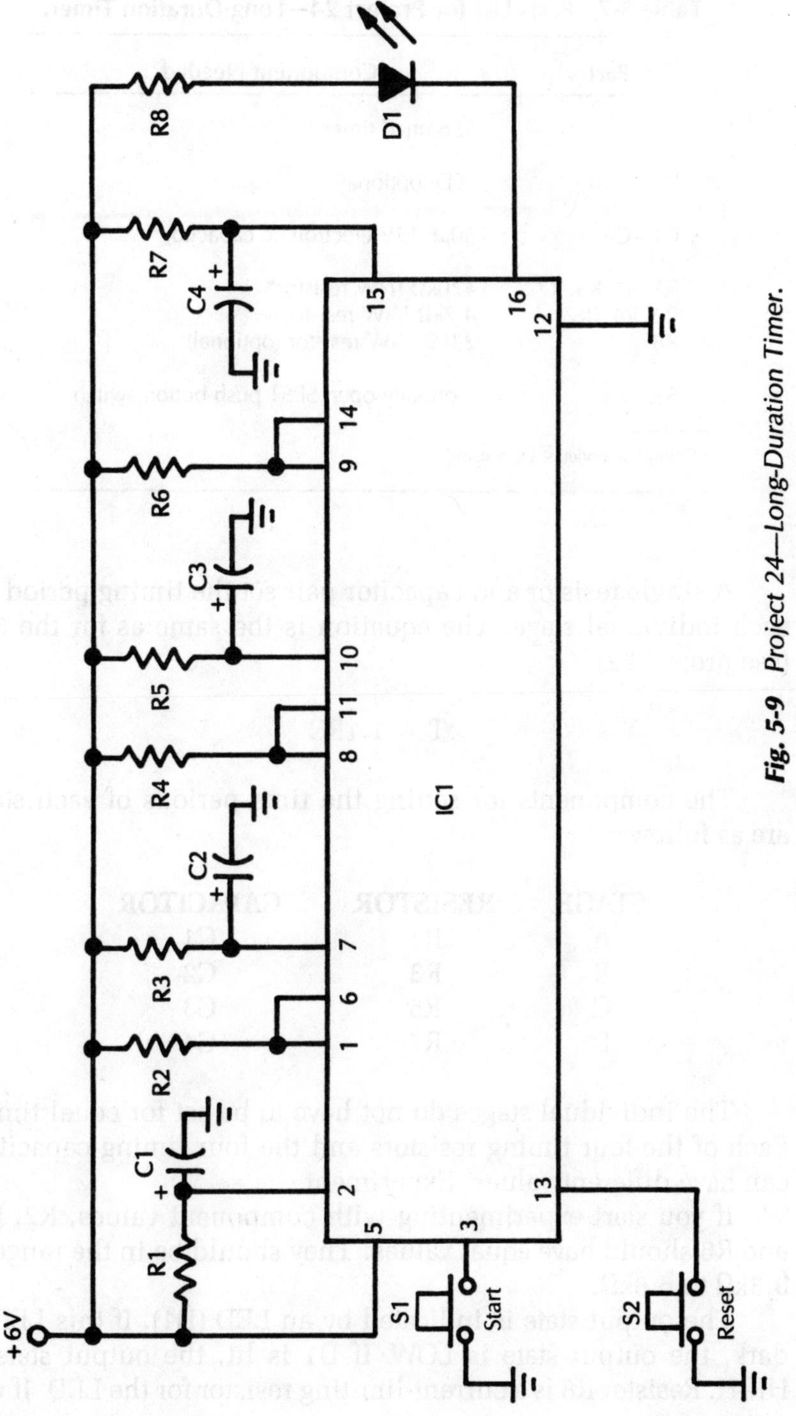


*Fig. 5-9 Project 24—Long-Duration Timer.*

**Table 5-7 Parts List for Project 24—Long-Duration Timer.**

| Part | Component Needed |
|---|---|
| IC1 | 558 quad timer |
| D1 | LED (optional) |
| C1 – C4 | 50$\mu$F 15V electrolytic capacitor* |
| R1, R3, R5, R7 | 470kΩ 1/4W resistor* |
| R2, R4, R6 | 4.7kΩ 1/4W resistor |
| R8 | 330Ω 1/4W resistor (optional) |
| S1, S2 | Normally open SPST push-button switch |

*Timing components; see text.

A single resistor and capacitor pair set the timing period for each individual stage. The equation is the same as for the 555 (see project 22):

$$T = 1.1RC$$

The components for setting the time periods of each stage are as follows:

| STAGE | RESISTOR | CAPACITOR |
|---|---|---|
| A | R1 | C1 |
| B | R3 | C2 |
| C | R5 | C3 |
| D | R7 | C4 |

The individual stages do not have to be set for equal times. Each of the four timing resistors and the four timing capacitors can have different values. Experiment.

If you start experimenting with component values, R2, R4, and R6 should have equal values. They should be in the range of 3.3kΩ to 6.8kΩ.

The output state is indicated by an LED (D1). If this LED is dark, the output state is LOW. If D1 is lit, the output state is HIGH. Resistor R8 is a current-limiting resistor for the LED. If you want to use your long-duration timer project to drive some other device, you can omit R8 and D1 from the circuit.

## PROJECT 25: SEQUENTIAL TIMER

In some applications, you might need different events to occur at different times. The solution to this kind of problem is a sequential timer. A suitable circuit is shown in Fig. 5-10, with a typical parts list appearing in Table 5-8.

This circuit is very similar to the long-duration timer circuit of project 24. The only real difference is that the outputs from each of the individual stages are being tapped off to drive other external devices.

A single resistor and capacitor pair set the timing period for each individual stage. The equation is the same as for the 555 (see project 22):

$$T = 1.1RC$$

The components for setting the time periods of each stage are as follows:

| STAGE | RESISTOR | CAPACITOR |
|---|---|---|
| A | R1 | C1 |
| B | R3 | C2 |
| C | R5 | C3 |
| D | R7 | C4 |

**Table 5-8 Parts List for Project 25—Sequential Timer.**

| Part | Component Needed |
|---|---|
| IC1 | 558 quad timer |
| C1 – C4 | 50$\mu$F 15V electrolytic capacitor* |
| R1, R3, R5, R7 | 470k$\Omega$ 1/4W resistor* |
| R2, R4, R6 | 4.7k$\Omega$ 1/4W resistor |
| S1, S2 | Normally open SPST push-button switch |

*Timing components; see text.

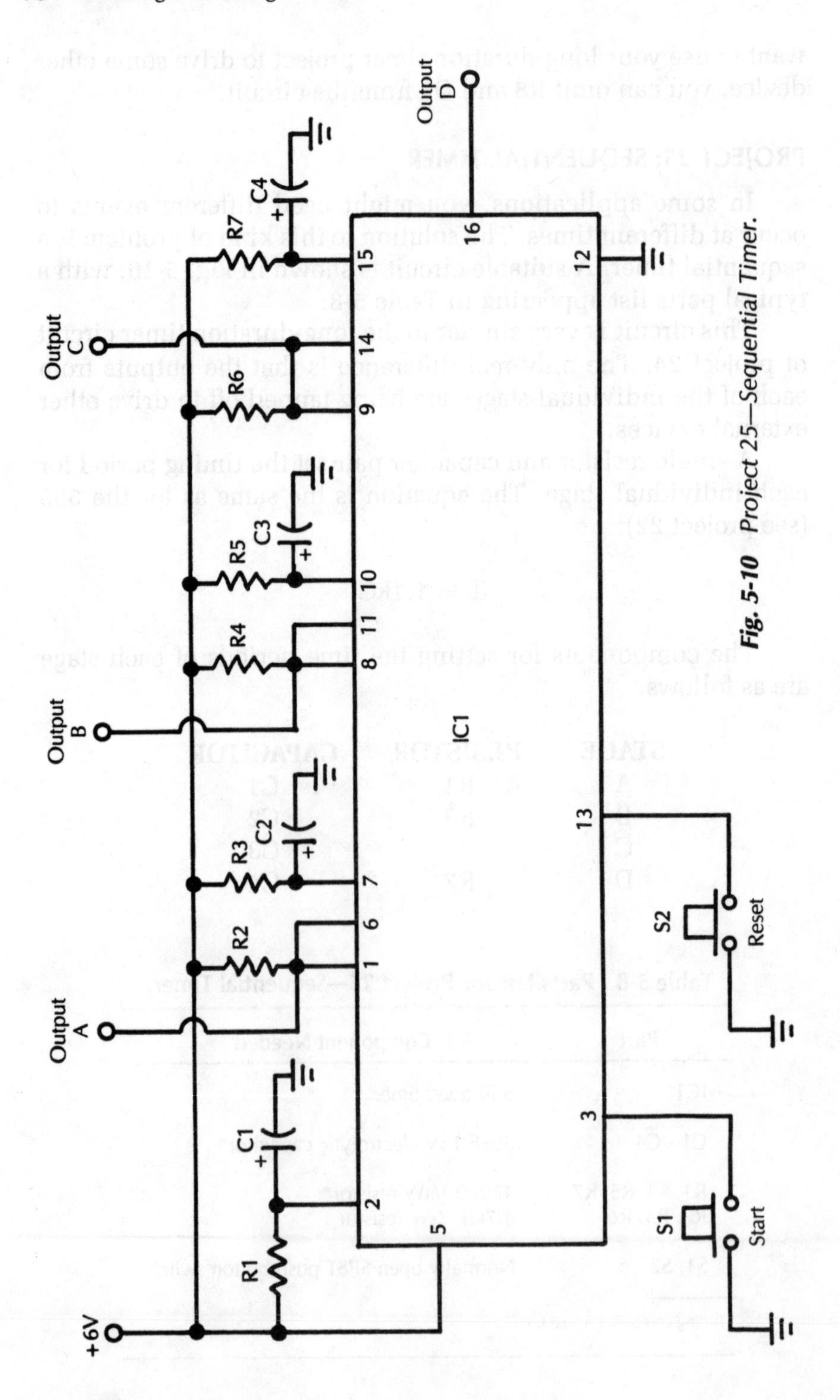


*Fig. 5-10 Project 25—Sequential Timer.*

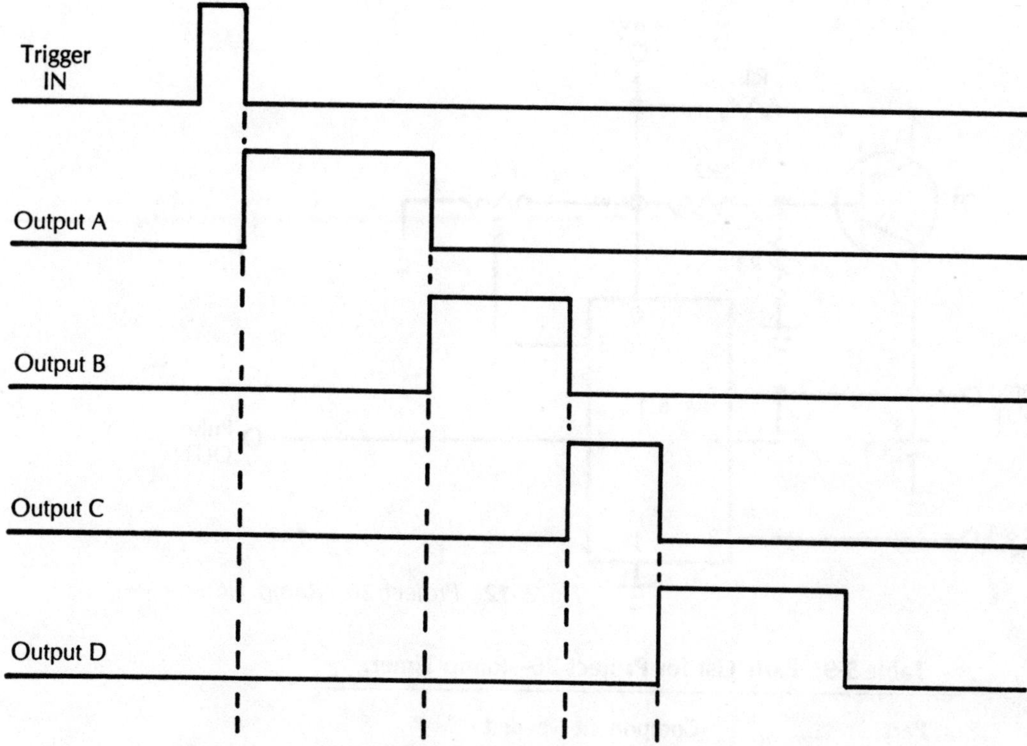


*Fig. 5-11* Some typical input and output signals for Fig. 5-10.

The individual stages do not have to be set for equal times. Each of the four timing resistors and the four timing capacitors can have different values. Experiment.

If you start experimenting with component values, R2, R4, and R6 should have equal values. They should be in the range of 3.3kΩ to 6.8kΩ. The operation of this circuit, including the input and output signals, is illustrated in Fig. 5-11.

## PROJECT 26: RAMP TIMER

In a few applications, you might not want a hard switching timer like the circuits of the last few projects. You might need a timed linear ramp. A circuit for accomplishing this is shown in Fig. 5-12, and a typical parts list is in Table 5-9.

The transistor (almost any PNP device will do) is a constant current source, which makes the charge across the capacitor linear. The timing period is set by the values of capacitor C1 and

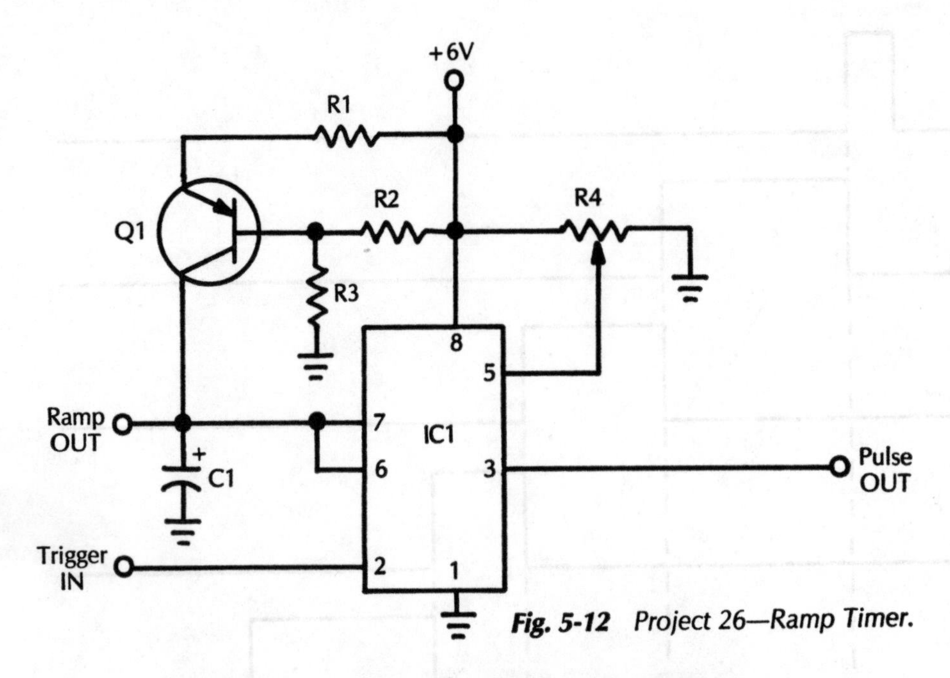


*Fig. 5-12 Project 26—Ramp Timer.*

**Table 5-9 Parts List for Project 26—Ramp Timer.**

| Part | Component Needed |
|---|---|
| IC1 | 555 (or 7555) timer |
| Q1 | PNP transistor to suit load (2N3906 or similar) |
| C1 | 0.1μF capacitor* |
| R1 | 47kΩ 1/4W resistor* |
| R2 | 4.7kΩ 1/4W resistor |
| R3 | 10kΩ 1/4W resistor |
| R4 | 10kΩ potentiometer |

*Experiment with other values.

resistor R1, using the same equation for any 555 monostable multivibrator circuit:

$$T = 1.1(R1)(C1)$$

Use potentiometer R4 to fine tune the linearity of the output ramp. Notice that the ordinary pulse (switching) output is simul-

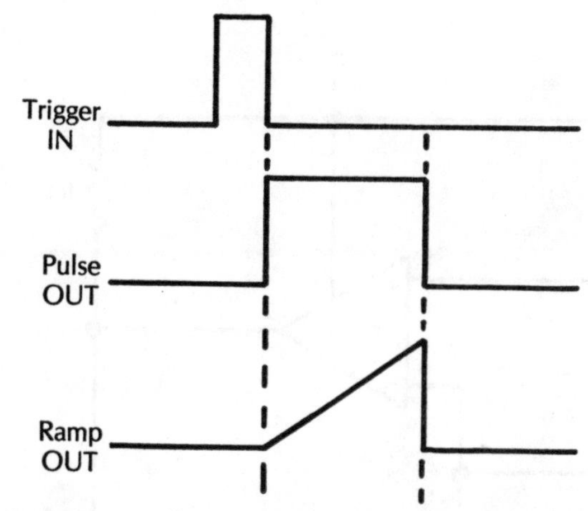


***Fig. 5-13*** *Input and output signals for Fig. 5-12.*

taneously available. Both the pulse and ramp outputs can be used to drive different external load devices at the same time. The input and output signals for this circuit are compared in Fig. 5-13.

## PROJECT 27: LIMIT COMPARATOR

If you ever need to know if a signal voltage is within a specific range, you can use the limit comparator circuit of Fig. 5-14. The parts list for this project is in Table 5-10.

This circuit features two comparator stages. The first stage (IC1) detects whether or not the input signal is above the upper threshold or limit. The second stage (IC2) detects whether or not the input signal is below the lower threshold or limit. The output goes HIGH only when the input voltage is above the lower threshold, but below the upper threshold.

The threshold voltages are set by the relative values of the resistive voltage-divider network made up of resistors R1, R2, and R3. These three resistors do not necessarily have to be equal. If they are equal (and assuming the supply voltage is +6V), the output will go HIGH when the input signal is between +2 and +4V. If the input signal is less than +2V or more than +4V, the output signal will go LOW.

For unequal upper and lower thresholds, just use Ohm's Law (R = E/I) to determine the required resistor values. As an exam-

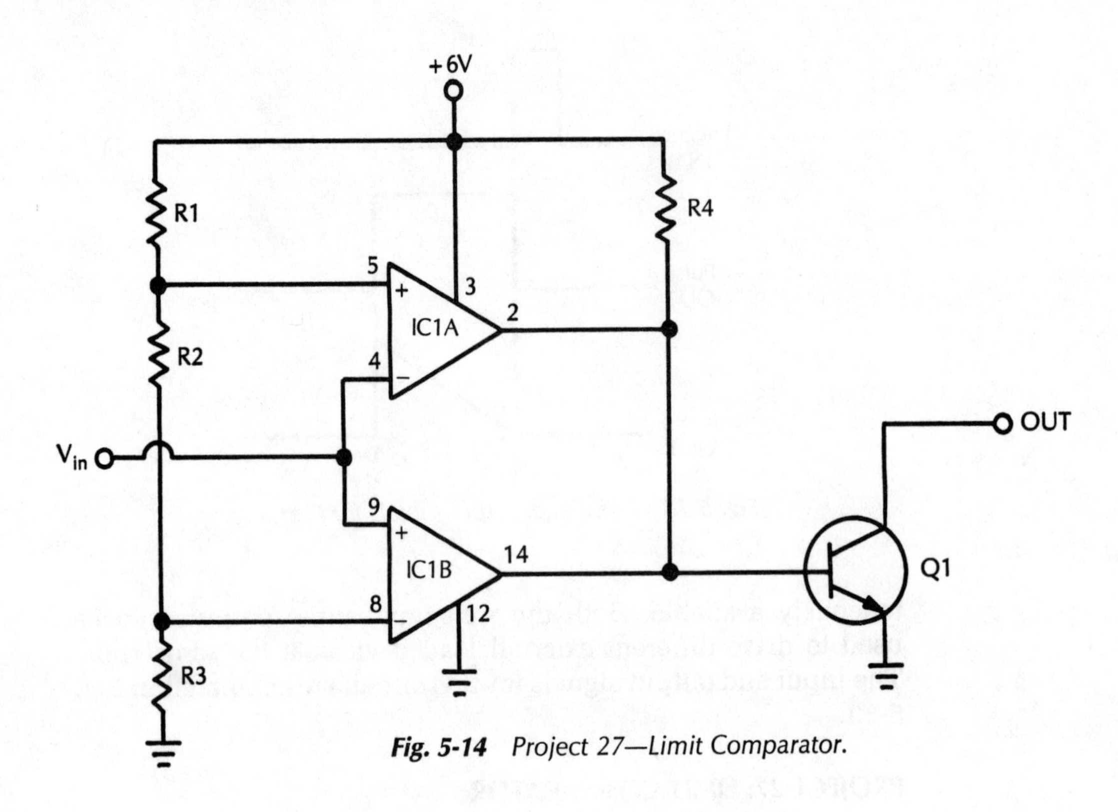


***Fig. 5-14*** *Project 27—Limit Comparator.*

**Table 5-10 Parts List for Project 27—Limit Comparator.**

| Part | Component Needed |
|---|---|
| IC1 | LM339 quad comparator |
| Q1 | NPN transistor to suit load (2N3904 or similar) |
| R1, R2, R3 | 3.3kΩ 1/4W resistor; see text. |
| R4 | 10kΩ 1/4W resistor |

ple, assume you are using the following resistances:

| | |
|---|---|
| R1 | 1kΩ |
| R2 | 2.2kΩ |
| R3 | 10kΩ |

The supply voltage, of course, is +6V.

The first step to calculating the upper and lower limits of the comparator window is to find the current through the resistance string. This is just a matter of using Ohm's Law. (Remember, resistances in series add.):

$$I = E/R$$

$$\begin{aligned} R &= 1000 + 2200 + 10000 \\ &= 13{,}200 \end{aligned}$$

$$\begin{aligned} I &= 6/13{,}200 \\ &= 0.0004545\text{A} \\ &= 0.4545\text{mA} \end{aligned}$$

Next, rearrange the Ohm's Law equation to find the voltage drop across each of the individual resistors:

$$E = IR$$

$$\begin{aligned} R1 &= 1000\Omega \\ E1 &= IR1 \\ &= 0.0004545 \times 1000 \\ &= 0.4545\text{V} \end{aligned}$$

$$\begin{aligned} R2 &= 2200\Omega \\ E2 &= IR2 \\ &= 0.0004545 \times 2200 \\ &= 0.9999\text{V} \end{aligned}$$

$$\begin{aligned} R3 &= 10000 \\ E3 &= IR3 \\ &= 0.0004545 \times 10000 \\ &= 4.545\text{V} \end{aligned}$$

To check your work, add the three voltage drops together to obtain the original supply voltage value (+6V):

$$\begin{aligned} V+ &= E1 + E2 + E3 \\ &= 0.4545 + 0.9999 + 4.545 \\ &= 5.9994 \end{aligned}$$

That is very close to the correct value. The small error is due to rounding in the earlier equations. Do not worry about such minor rounding errors.

The lower limit of the on range is set by the voltage drop across resistor R3, or 4.545V above ground in our example. The upper limit of the on range is set by the supply voltage, minus the voltage drop across resistor R1. In the example, this works out to:

$$\begin{aligned} & V+ - E1 \\ &= 6 - 0.4545 \\ &= 5.5455\text{V} \end{aligned}$$

Using the component values in this example, the limit comparator output will go HIGH whenever the input voltage is between +4.545V and +5.5455V.

Resistor R2 sets the range width between the upper and lower limits. This range width is equal to the voltage drop across this resistor. In the example, the window width is just under 1V.

Breadboarding and experimenting with this circuit is strongly recommended. For maximum versatility in your project, you might want to consider using potentiometers for the voltage divider resistances (R1, R2, and R3).

## PROJECT 28: SWITCH DEBOUNCER

Nothing is perfect, and that certainly includes everything you work with in electronics. A mechanical switch certainly seems simple enough. How complicated could it be? For most applications (especially analog applications), a mechanical switch is perfectly straightforward enough. Either the contacts are open or they are closed. Right?

But a lot depends on how fast you are looking at the switch. When any mechanical switch is moved from one position to another, the contacts will inevitably *bounce* several times before settling into their new resting position. Instead of cleanly switching from on to off, as you would assume, the switch will actually oscillate:

On - Off - On - Off - On - Off - On - Off -

for a fraction of a second, before finally stopping in a full off position. This bounce is illustrated in Fig. 5-15.

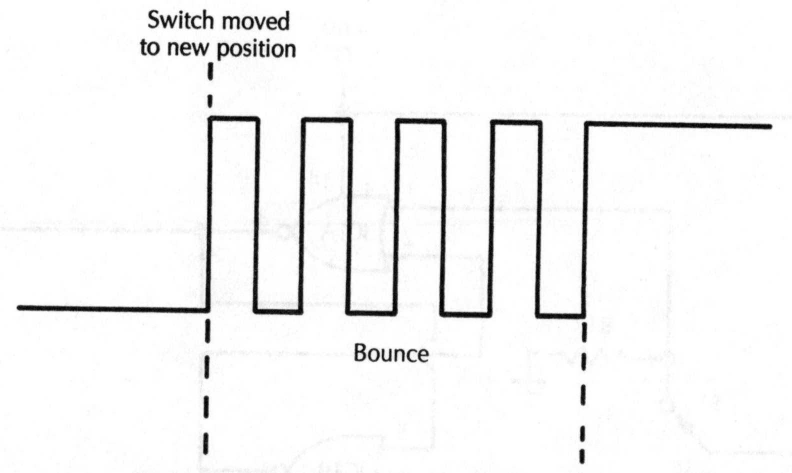


*Fig. 5-15* Mechanical switches tend to bounce.

In a great many applications, including most analog circuits, this switch bouncing will not make the slightest bit of difference (unless it is extremely severe). But with digital circuitry, it is another story. Digital circuits can usually respond to very brief input pulses—that is the way they are designed. The circuit cannot tell the difference between contact bounce and a true change in the switch setting. To understand why this could be a problem, consider as an example an application in which a mechanical switch is used to increment a digital counter. The counter is supposed to keep track of how many times the switch is closed. (Assume a momentary action, normally-open switch.) If the counter accepts each contact bounce as a new switch closure, you will obviously end up with a wildly inaccurate count. Instead of:

1 - 2 - 3 - 4 - 5 - 6 - 7 - 8 -

you might get something like:

1 - 5 - 8 - 11 - 12 - 17 - 21 - 26

Quite plainly, this would be absolutely useless, except perhaps as a crude random-number generator.

The solution to such problems is to use a circuit known as a *switch debouncer.* This is essentially nothing more than a monostable multivibrator with a rather brief time period. The

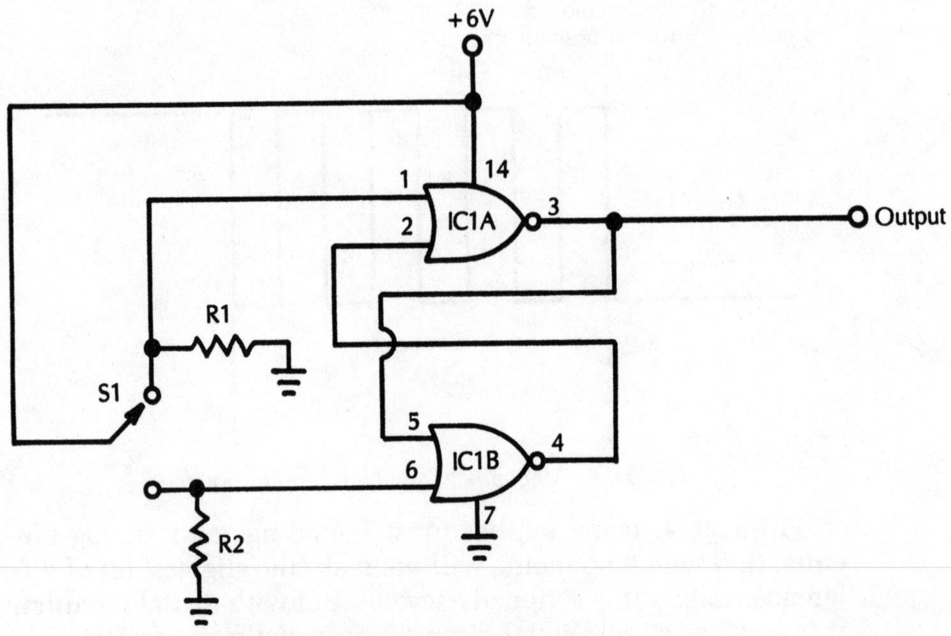


*Fig. 5-16 Project 28—Switch Debouncer.*

**Table 5-11 Parts List for Project 28—Switch Debouncer.**

| Part | Component Needed |
|---|---|
| IC1 | CD4001 quad NOR gate |
| R1, R2 | 100kΩ 1/4W resistor |
| S1 | SPDT switch |

first time the switch contacts close (the first bounce), the multivibrator is triggered. Its time constant is selected to last slightly longer than the bounce period. The output circuit sees a single, clean pulse, rather than a batch of irregular, jagged bounce pulses. (Refer back to Fig. 5-15.)

The schematic diagram for a switch debouncer project appears in Fig. 5-16, and the parts list for this project is in Table 5-11. Nothing is critical here. You might want to experiment with other resistor values. Both resistors should have identical values. The resistance will have an effect on the timing period of the switch-debouncer circuit. For most switch-debouncing applications, 100kΩ or so is just about ideal.

# 6 ❖ Digital Demonstration Circuits

IT IS GETTING HARDER AND HARDER FOR ANYONE INVOLVED IN electronics to avoid digital circuitry. A lot of people fear that digital logic might be too difficult to understand. Terms like *Boolean algebra* are quite intimidating. Actually, digital electronics is relatively simple. Everything is ultimately built up from a few simple circuit units known as gates.

The projects in this chapter are not practical, finished projects like most of those in the preceeding chapters. Instead, these projects are demonstration circuits to help you understand how various digital devices work. These simple circuits can be combined to perform more complex functions. Some of the projects in this book would make a good start on a strong science fair project.

## DIGITAL SIGNALS

An analog signal can take on any of an infinite number of values. Digital signals, on the other hand, can only have one of two possible levels, or states. In this book, the two states are called LOW and HIGH. Other names are sometimes used:

| LOW | HIGH |
|---|---|
| 0 | 1 |
| OFF | ON |
| NO | YES |

Digital circuitry is said to operate in binary. *Binary* is a number system with just two possible digits: 0 (zero) and 1 (one). You are more familiar with the decimal system, which has ten possible digits:

0 1 2 3 4 5 6 7 8 9

To express a number greater than 9 in the decimal system, you need to add an extra digit column. For example:

13
47
198

Similarly, in the binary system a new column is needed whenever the value in a given column exceeds one:

10
1010
11101

Here is a comparison of the binary and decimal numbering systems:

| BINARY | DECIMAL |
|---|---|
| 0000 | 0 |
| 0001 | 1 |
| 0010 | 2 |
| 0011 | 3 |
| 0100 | 4 |
| 0101 | 5 |
| 0110 | 6 |
| 0111 | 7 |
| 1000 | 8 |
| 1001 | 9 |
| 1010 | 10 |
| 1011 | 11 |
| 1100 | 12 |
| 1101 | 13 |
| 1110 | 14 |
| 1111 | 15 |
| 10000 | 16 |

For most elementary hobby work, you will not need to do any conversion between the binary and decimal numbering systems. But you need to be able to recognize binary numbers and have at least a rough understanding of how the binary number system works.

The binary system might seem very unwieldy and awkward. And it is, for us humans. But for electronic circuits, it is the easiest and most natural possible numbering system. The two possible digits can easily and unambiguously be represented electrically. LOW is a low-voltage signal (close to ground potential), and HIGH is a higher voltage signal (close to the positive supply voltage). You do not need to worry about any in-between values.

## BASICS OF DIGITAL GATES

A *digital gate* is a circuit that accepts one or more binary, (or digital) signals, and puts out an output signal in response to the input(s) according to a specific pattern. A written summary of the input/output pattern for a specific gating circuit is called a *truth table*.

The simplest possible digital gate is the *buffer*. It has one input and one output, as shown in Fig. 6-1. The output is always identical to the input. The truth table looks like this:

| INPUT | OUTPUT |
|---|---|
| 0 | 0 |
| 1 | 1 |

This might seem like an exercise in uselessness. The buffer gate does not change the signal at all. The sole function of a buffer gate is implied by the name. It is similar in function to an analog buffer amplifier. A buffer gate is used so that an output signal from one gate can drive more inputs of other gates. Buffer gates are not too commonly used.

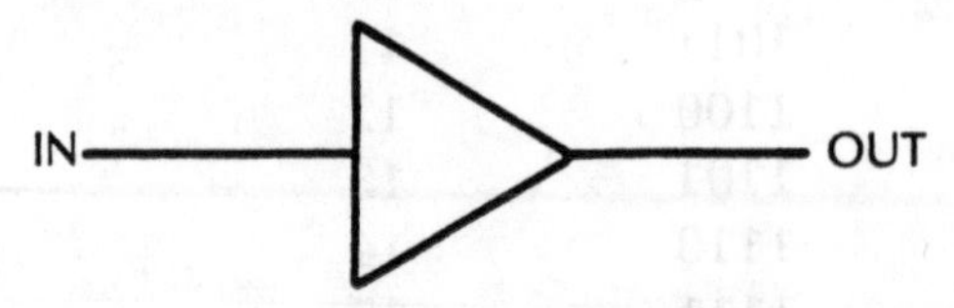


***Fig. 6-1** The simplest possible digital gate is the buffer.*

There are three main gate types that are widely used. Ultimately, all digital circuits are made up of various combinations of these three types of gates. They are the inverter, the AND gate, and the OR gate.

The *inverter*, like the buffer, has just a single input and a single output. The schematic symbol for an inverter is shown in Fig. 6-2. A small circle, like the one at the output end of this symbol, is used to indicate an inversion function.

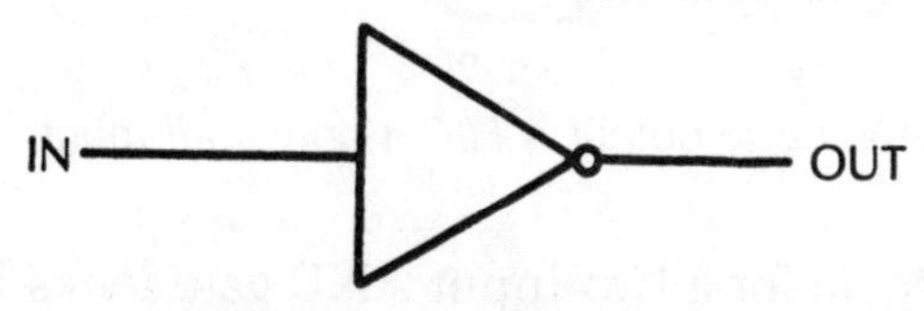


***Fig. 6-2*** *An inverter has opposite input and output signals.*

The name is self-explanatory. The inverter inverts, or reverses, the input signal at the output. The output is always at the opposite state from the input signal. Inverters are sometimes called *NOT* gates. The truth table for an inverter looks like this:

| INPUT | OUTPUT |
|---|---|
| 0 | 1 |
| 1 | 0 |

More complex gates, such as the AND gate and the OR gate, have two (or more) inputs and a single output. A two-input digital gate can accept four distinct input combinations:

| | |
|---|---|
| 0 | 0 |
| 0 | 1 |
| 1 | 0 |
| 1 | 1 |

An *AND* gate, shown in Fig. 6-3, has a HIGH output if, and only if, all inputs are HIGH. If any one (or more) of the inputs is LOW, then the output will be LOW too.

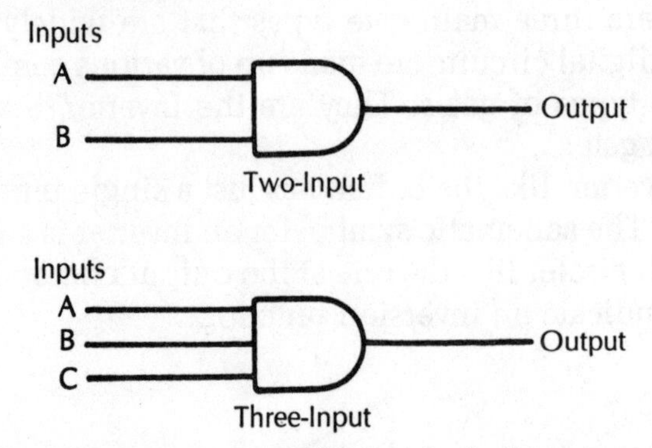


***Fig. 6-3*** *An AND gate output is HIGH only if all inputs are HIGH.*

The truth table for a two-input AND gate looks like this:

| INPUTS A B | OUTPUT |
|---|---|
| 0 0 | 0 |
| 0 1 | 0 |
| 1 0 | 0 |
| 1 1 | 1 |

An AND gate with more than two inputs works in the same way. If at least one of the inputs is LOW, then the output must be LOW also. The output is HIGH if (and only if) all of the inputs are HIGH. For example, for a three-input AND gate:

| INPUTS A B C | OUTPUT |
|---|---|
| 0 0 0 | 0 |
| 0 0 1 | 0 |
| 0 1 0 | 0 |
| 0 1 1 | 0 |
| 1 0 0 | 0 |
| 1 0 1 | 0 |
| 1 1 0 | 0 |
| 1 1 1 | 1 |

An *OR* gate is illustrated in Fig. 6-4. For an OR gate, the output is HIGH if at least one input is HIGH. The output is LOW if (and only if) all inputs are LOW. For a two-input gate, you can

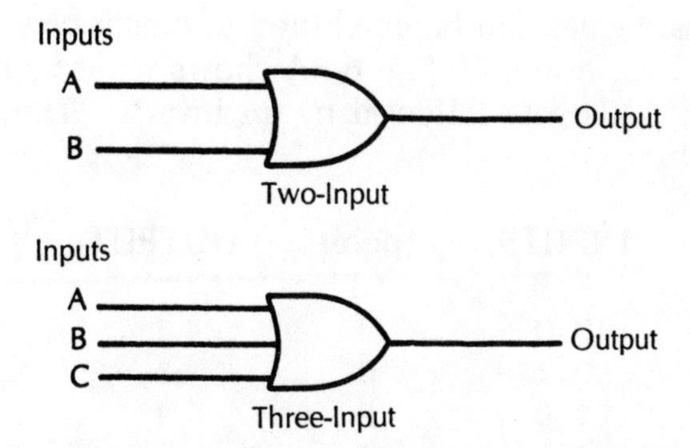


*Fig. 6-4 An OR gate output is HIGH if at least one input is HIGH.*

remember, the output is HIGH if (and only if) either input A or input B is HIGH. The truth table looks like this:

| INPUTS A | INPUTS B | OUTPUT |
|---|---|---|
| 0 | 0 | 0 |
| 0 | 1 | 1 |
| 1 | 0 | 1 |
| 1 | 1 | 1 |

Like the AND gate, the OR gate can have more than two inputs. The multi-input versions work in a similar manner.

There is a special variant of the basic OR gate, known as the Exclusive-OR, or *XOR* gate. The XOR symbol appears in Fig. 6-5.

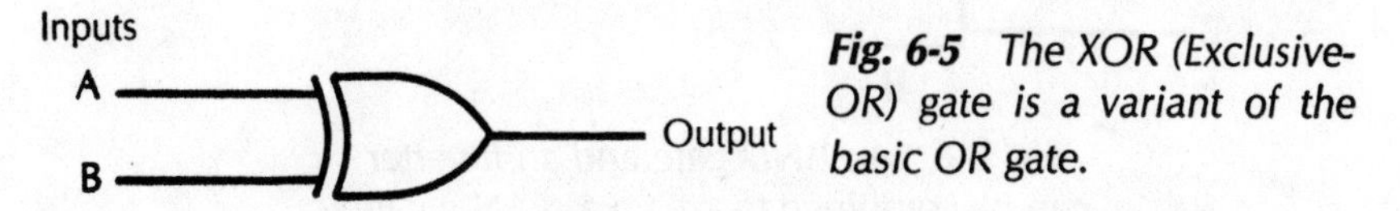


*Fig. 6-5 The XOR (Exclusive-OR) gate is a variant of the basic OR gate.*

The output is HIGH if (and only if) input A is HIGH or input B is HIGH, but not both. The XOR gate might be considered a difference detector. The output is HIGH if the inputs are not alike. The truth table looks like this:

| INPUTS A | INPUTS B | OUTPUT |
|---|---|---|
| 0 | 0 | 0 |
| 0 | 1 | 1 |
| 1 | 0 | 1 |
| 1 | 1 | 0 |

The basic gates can be combined to create new patterns or truth tables. For example, Fig. 6-6A shows what happens when we have an AND gate followed by an inverter. The truth table looks like this:

| INPUTS A B | (point C) | OUTPUT |
|---|---|---|
| 0 0 | 0 | 1 |
| 0 1 | 0 | 1 |
| 1 0 | 0 | 1 |
| 1 1 | 1 | 0 |

The output pattern is just the opposite of an AND gate. This is called a *NAND* gate (for Not AND). This function is so widely used, dedicated NAND gate devices are commonly available. The schematic symbol for a NAND gate appears in Fig. 6-6B.

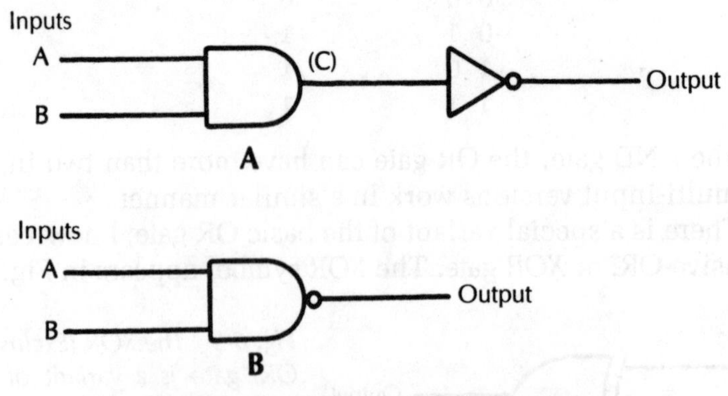


***Fig. 6-6*** *An AND gate and an inverter can be combined to create a NAND gate.*

Similarly, a NOR (Not OR) gate is an OR gate followed by an inverter, as shown in Fig. 6-7. The truth table is as follows:

| INPUTS A B | (point C) | OUTPUT |
|---|---|---|
| 0 0 | 0 | 1 |
| 0 1 | 1 | 0 |
| 1 0 | 1 | 0 |
| 1 1 | 1 | 0 |

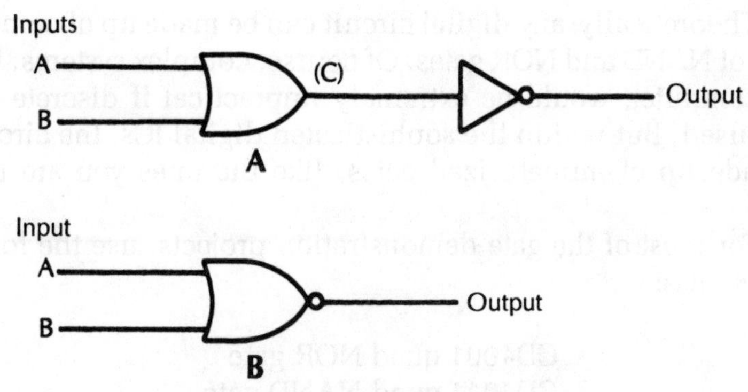


*Fig. 6-7* *An OR gate and an inverter can be combined to create a NOR gate.*

For various technical reasons, it is more practical to make NAND and NOR gates. These are the most widely used digital gates. They are the most readily available and the least expensive. If an AND or OR function is needed, adding another inverter at the output will do the trick.

By shorting together the inputs of either a NAND gate or a NOR gate, an inverter can be simulated. For a NAND gate:

| INPUTS A B | OUTPUT |
|---|---|
| 0 0 | 1 |
| 0 1 | NOT POSSIBLE |
| 1 0 | NOT POSSIBLE |
| 1 1 | 0 |

If the inputs are shorted together, it is impossible for the two inputs to take on differing values. They must be identical.

Using a NOR gate with the inputs shorted together, you get the same pattern:

| INPUTS A B | OUTPUT |
|---|---|
| 0 0 | 1 |
| 0 1 | NOT POSSIBLE |
| 1 0 | NOT POSSIBLE |
| 1 1 | 0 |

Theoretically, any digital circuit can be made up of combinations of NAND and NOR gates. Of course, complex systems, like a full computer, would be extremely impractical if discrete gates were used. But within the sophisticated digital ICs, the circuitry is made up of miniaturized gates, like the ones you are using here.

For most of the gate demonstration projects, use the following two ICs:

CD4001 quad NOR gate
CD4011 quad NAND gate

You can achieve all simple gating functions with these chips.

## PROJECT 29: NAND GATE DEMONSTRATOR

The circuit shown in Fig. 6-8 demonstrates the operation of a digital NAND gate. The parts list for this project is in Table 6-1.

Each input is manually controlled by an ordinary SPST switch (S1 and S2). Consider only switch S1 in this discussion. Both input switching networks work in exactly the same way.

When the switch is open, the gate input is grounded through resistor R1. This appears to the gate as a logic LOW signal. The exact value of this resistor is not critical, but it should be fairly

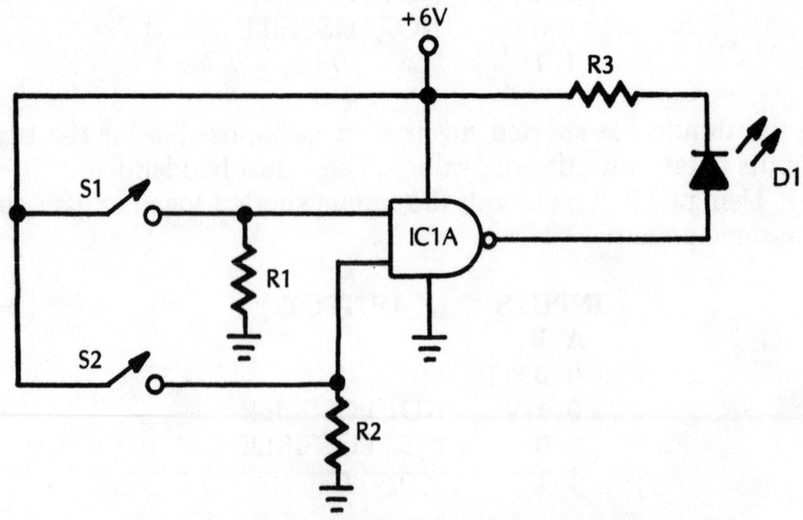


*Fig. 6-8* *Project 29—NAND Gate Demonstrator.*

**Table 6-1 Parts List for Project 29—NAND Gate Demonstrator.**

| Part | Component Needed |
|---|---|
| IC1 | CD4011 quad NAND gate |
| D1 | LED |
| R1, R2 | 1MΩ 1/4W resistor |
| R3 | 330Ω 1/4W resistor |
| S1, S2 | SPST switch |

large to prevent excessive current drain when the switch is closed.

When the switch is closed, the gate input is shorted to the positive supply voltage. This looks like a logic HIGH signal to the gate. When working with this circuit (and the other digital demonstration projects in this chapter), just remember that an open switch is LOW, and a closed switch is HIGH.

The output state is indicated by the LED (D1). Resistor R3 is a current-dropping resistor to prevent the LED from drawing excessive current and burning itself out. The exact value is not critical, although it will affect how brightly the lit LED glows. The value of R3 should be between 200 and 1000Ω.

Experiment with various settings of the input switches. There are only four possible combinations. Does the output state, as indicated by the LED, correspond to what is predicted by the truth table for a NAND gate?

| INPUTS A | INPUTS B | OUTPUT |
|---|---|---|
| 0 | 0 | 1 |
| 0 | 1 | 1 |
| 1 | 0 | 1 |
| 1 | 1 | 0 |

## PROJECT 30: INVERTER DEMONSTRATOR

By wiring the two inputs of a NAND gate together, as shown in Fig. 6-9, you create an inverter demonstration circuit. The parts list for this project is in Table 6-2. Notice that only a single

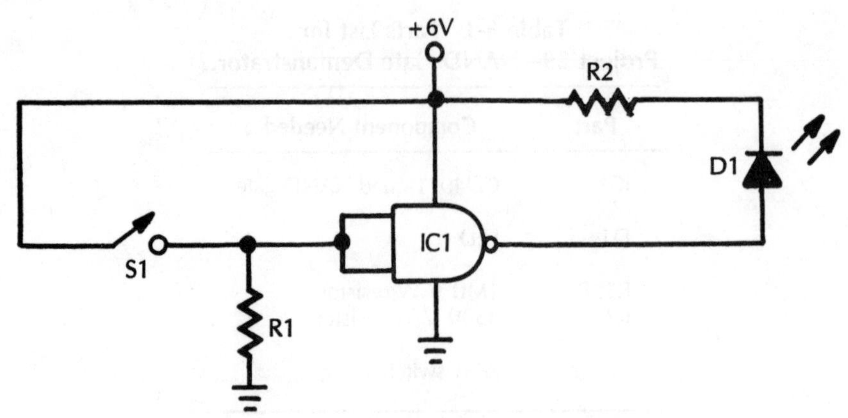


*Fig. 6-9* *Project 30—Inverter Demonstrator.*

**Table 6-2 Parts List for Project 30—Inverter Demonstrator.**

| Part | Component Needed |
|---|---|
| IC1 | CD4011 quad NAND gate |
| D1 | LED |
| R1 | 1MΩ 1/4W resistor |
| R3 | 330Ω 1/4W resistor |
| S1 | SPST switch |

input-switch network (S1 and R1) is needed for this project. An inverter is a single input/single output device. Otherwise, this circuit functions just like the NAND gate demonstrator of project 29.

When the switch is open, the gate input is grounded through resistor R1. This appears to the gate as a logic LOW signal.

When the switch is closed, the gate input is shorted to the positive supply voltage. This looks like a logic HIGH signal to the gate. Just remember that an open switch is LOW, and a closed switch is HIGH. The output state is indicated by the LED (D1). Resistor R3 is a current-dropping resistor for the LED.

Try the switch (S1) in both positions. The output state, as indicated by the LED, should always be the opposite of the input signal:

| SWITCH POSITION | INPUT | OUTPUT |
|---|---|---|
| OPEN | 0 | 1 |
| CLOSED | 1 | 0 |

## PROJECT 31: BUFFER-GATE DEMONSTRATOR

A buffer-gate can be simulated by two inverter stages in series, as shown in Fig. 6-10, and a parts list for this project is in Table 6-3. Like the inverter circuit of project 30, this is a single input/single output gate. Only one input switch network (S1 and R1) is required.

When the switch is open, the input signal is LOW. IC1A inverts this signal to a HIGH level. IC1B then re-inverts the HIGH signal back to a LOW signal.

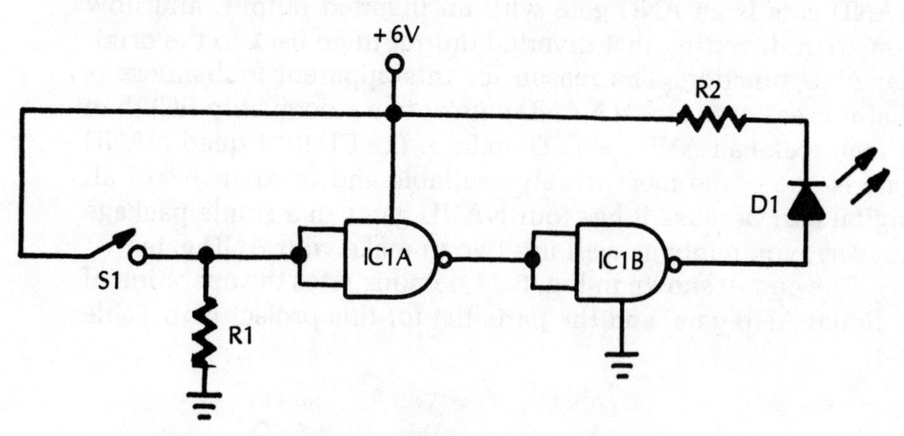


*Fig. 6-10 Project 31—Buffer-Gate Demonstrator.*

**Table 6-3 Parts List for Project 31—Buffer-Gate Demonstrator.**

| Part | Component Needed |
|---|---|
| IC1 | CD4011 quad NAND gate |
| D1 | LED |
| R1 | 1MΩ 1/4W resistor |
| R3 | 330Ω 1/4W resistor |
| S1 | SPST switch |

Similarly, when the switch is closed, the input signal is HIGH. IC1A inverts this signal to a LOW level. IC1B then re-inverts the LOW signal back to its original HIGH state.

The output signal, as indicated by the LED (D1) should always be the same as the input signal as set by switch S1:

| SWITCH POSITION | INPUT | OUTPUT |
|---|---|---|
| OPEN | 0 | 0 |
| CLOSED | 1 | 1 |

## PROJECT 32: AND GATE DEMONSTRATOR

An AND gate can be easily simulated by inverting the output of a NAND gate. At first glance, this seems a little silly. A NAND gate is an AND gate with an inverted output, and now you are re-inverting that inverted output to go back to the original AND function. The reason for this apparent foolishness is that it is easier to etch a NAND gate onto a silicon chip (IC) than it is to make an AND gate. Therefore, the CD4011 quad NAND gate is one of the most widely available and inexpensive of all digital ICs. Because it has four NAND gates in a single package anyway, you might as well use two to make your AND gate.

The circuit shown in Fig. 6-11 demonstrates the operation of a digital AND gate, and the parts list for this project is in Table 6-4.

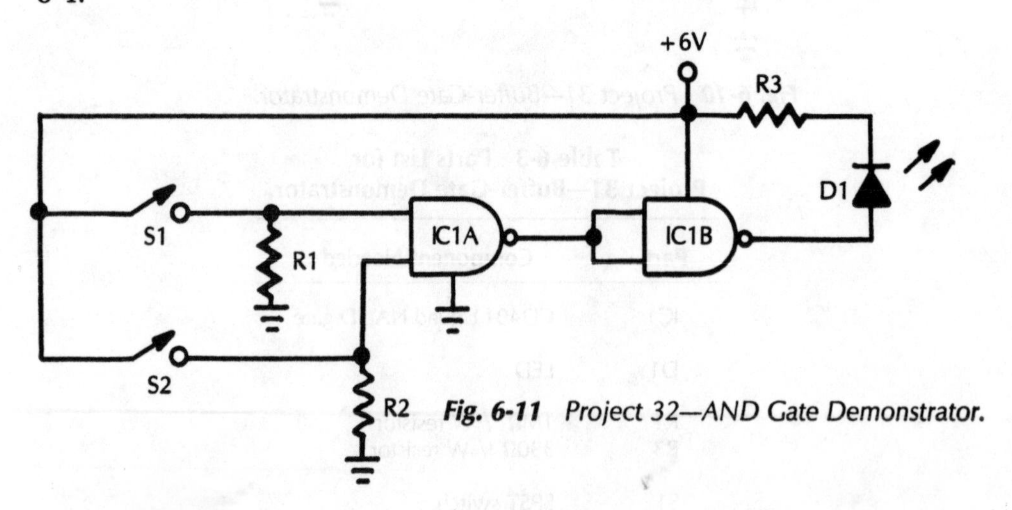


*Fig. 6-11 Project 32—AND Gate Demonstrator.*

**Table 6-4 Parts List for Project 32—AND Gate Demonstrator.**

| Part | Component Needed |
|---|---|
| IC1 | CD4011 quad NAND gate |
| D1 | LED |
| R1, R2 | 1MΩ 1/4W resistor |
| R3 | 330Ω 1/4W resistor |
| S1, S2 | SPST switch |

Each input is manually controlled by an ordinary SPST switch (S1 and S2). Consider only switch S1 in this discussion. Both input switching networks work in exactly the same way.

When the switch is open, the gate input is grounded through resistor R1. This appears to the gate as a logic LOW signal.

When the switch is closed, the gate input is shorted to the positive supply voltage. This looks like a logic HIGH signal to the gate. Remember that an open switch is LOW, and a closed switch is HIGH.

The output state is indicated by the LED (D1). Resistor R3 is a current-dropping resistor to prevent the LED from drawing excessive current and burning itself out. The exact value is not critical, although it will affect how brightly the lit LED glows. The value of R3 should be between 200 and 1000Ω.

Experiment with various settings of the input switches. There are only four possible combinations. Does the output state, as indicated by the LED, correspond to what is predicted by the truth table for an AND gate?

| INPUTS A | INPUTS B | OUTPUT |
|---|---|---|
| 0 | 0 | 0 |
| 0 | 1 | 0 |
| 1 | 0 | 0 |
| 1 | 1 | 1 |

## PROJECT 33: NOR GATE DEMONSTRATOR

The circuit shown in Fig. 6-12 demonstrates the operation of a digital NOR (Not OR) gate, and the parts list for this project appears as Table 6-5.

Each input is manually controlled by an ordinary SPST switch (S1 and S2). These input switches function in exactly the same way as in the NAND gate demonstrator circuit of project 29.

Experiment with various settings of the input switches. There are only four possible combinations. Does the output state, as indicated by the LED, correspond to what is predicted by the truth table for a NOR gate?

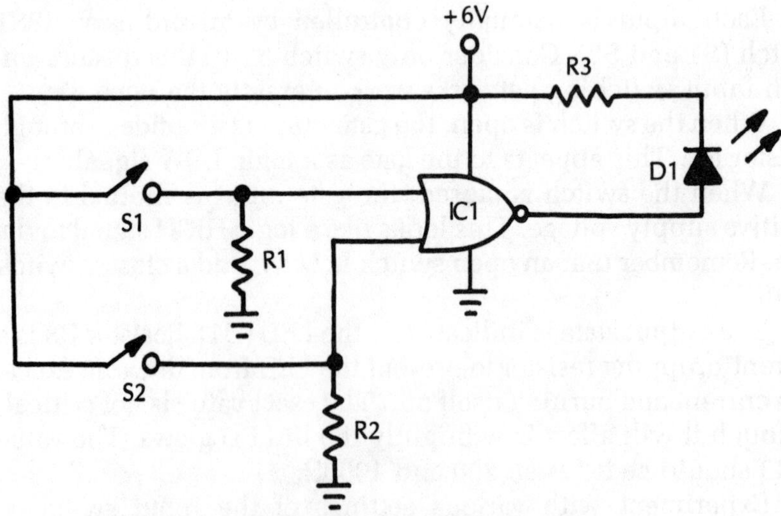


*Fig. 6-12 Project 33—NOR Gate Demonstrator.*

**Table 6-5 Parts List for Project 33—NOR Gate Demonstrator.**

| Part | Component Needed |
|---|---|
| IC1 | CD4001 quad NOR gate |
| D1 | LED |
| R1, R2 | 1MΩ 1/4W resistor |
| R3 | 330Ω 1/4W resistor |
| S1, S2 | SPST switch |

| INPUTS | | OUTPUT |
|---|---|---|
| A | B | |
| 0 | 0 | 1 |
| 0 | 1 | 0 |
| 1 | 0 | 0 |
| 1 | 1 | 0 |

## PROJECT 34: OR GATE DEMONSTRATOR

An OR gate can be easily simulated by inverting the output of a NOR gate. This may seem like an exercise in pointlessness, because a NOR gate is nothing more than an OR gate with an inverted output, and now you are re-inverting that inverted output to go back to the original OR gate function.

It is not as silly as it all seems, however. For various technical reasons, NOR gates are easier to make in IC form than OR gates. This means that the CD4001 quad NOR gate is one of the most widely available and inexpensive of all digital ICs. Because it has four NOR gates in a single package it is no problem to use two to make the OR gate simulator. It does not increase the parts count of the project by even one additional component.

The schematic diagram for the OR gate demonstrator circuit is illustrated in Fig. 6-13, and the parts list is in Table 6-6.

There is nothing new to say about this circuit. It works in the same basic fashion as projects 29–33. The only difference is that

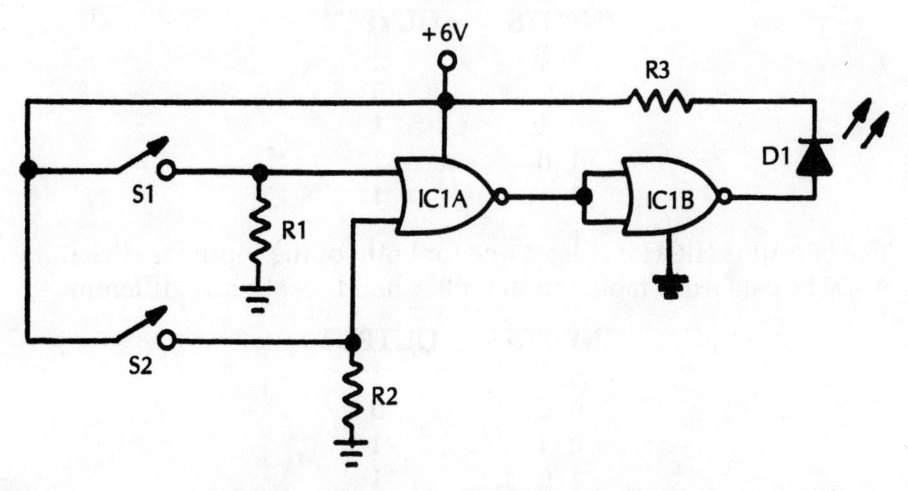


*Fig. 6-13* Project 34—OR Gate Demonstrator.

**Table 6-6 Parts List for Project 34—OR Gate Demonstrator.**

| Part | Component Needed |
|---|---|
| IC1 | CD4001 quad NOR gate |
| D1 | LED |
| R1, R2 | 1MΩ 1/4W resistor |
| R3 | 330Ω 1/4W resistor |
| S1, S2 | SPST switch |

the LED (D1) should indicate the output pattern predicted by the truth table for an OR gate:

| INPUTS A B | OUTPUT |
|---|---|
| 0 0 | 0 |
| 0 1 | 1 |
| 1 0 | 1 |
| 1 1 | 1 |

## PROJECT 35: EXCLUSIVE-OR GATE DEMONSTRATOR

The XOR gate is a special variation on the OR gate. The truth table for an ordinary OR gate looks like this:

| INPUTS A B | OUTPUT |
|---|---|
| 0 0 | 0 |
| 0 1 | 1 |
| 1 0 | 1 |
| 1 1 | 1 |

The output is HIGH if at least one (or both) of the inputs is HIGH. An XOR gate truth table, on the other hand, is slightly different:

| INPUTS A B | OUTPUT |
|---|---|
| 0 0 | 0 |
| 0 1 | 1 |
| 1 0 | 1 |
| 1 1 | 0 |

The output is HIGH only if one (but not both) of the inputs is HIGH. Another way of looking at this gate is to say its output is HIGH if the two inputs have opposite values or LOW if the two inputs are the same.

Dedicated XOR gate chips are available, but for this demonstration circuit (shown in Fig. 6-14) you will build an exclusive-OR gate using NOR gates. This project demonstrates how any gating circuit can be created from the basic types of gates.

Six NOR gate stages are used in this circuit. Because a CD4001 has four stages, two ICs will be required to construct this project. The complete parts list appears in Table 6-7.

Overall, the operation of this circuit is the same as for all the gate demonstration circuits presented so far in this chapter. The digital signals at various points in the circuit are summarized in Table 6-8.

## PROJECT 36: MAJORITY-LOGIC DEMONSTRATOR

In most digital-gating applications, the output state is determined by very specific patterns of input states. In some cases, however, exact patterns of individual inputs being at specific states are not as important as broader, more generalized patterns.

As an example, consider a voting-machine network. Multiple inputs are all given equal weightings. Who votes for whom is not important in determining the outcome of the election. The winning candidate is the one with the greatest number of votes.

In terms of a digital gate, the state of input C, for example, is not particularly important in itself. Which input state is dominant over the entire set of all inputs is important. That is, which logic state gets the greatest number of "votes." In other words, rather than direct, fixed-gating logic, as with the projects so far in this chapter, you need a system of "majority rule." Not surprisingly, gating circuits that work along these lines are known as *majority-logic* circuits.

Dedicated majority-logic chips are available. Unfortunately, they can often be fairly difficult for the hobbyist to locate. It is also more informative to use standard gates, so you can follow what is happening in the circuit step by step. Incidentally, this project is particularly well suited to use in science fairs.

A majority-logic gating circuit allows the various inputs to vote on the desired output state. In a noninverted circuit, if the

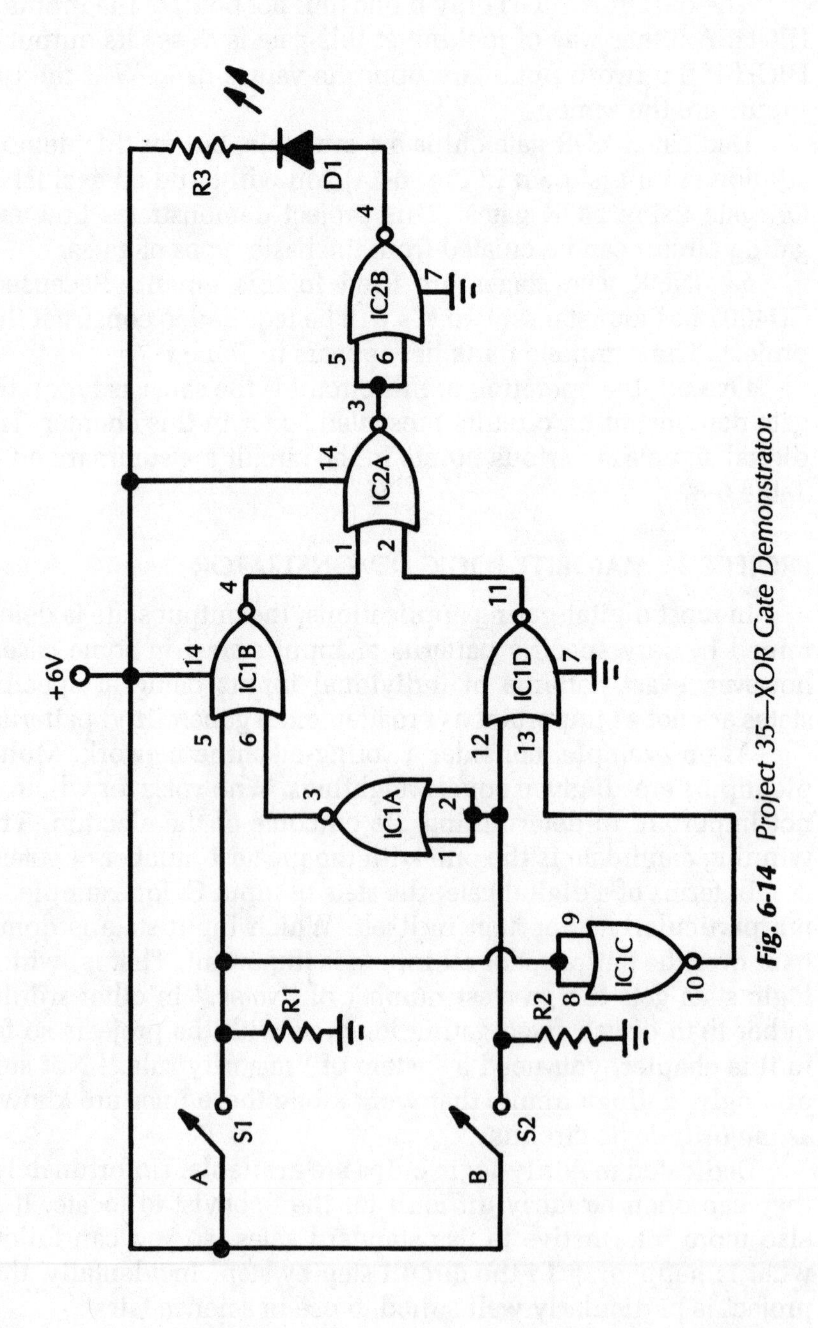


*Fig. 6-14* Project 35—XOR Gate Demonstrator.

**Table 6-7 Parts List for Project 35—XOR Gate Demonstrator.**

| Part | Component Needed |
|---|---|
| IC1, IC2 | CD4001 quad NOR gate |
| D1 | LED |
| R1, R2 | 1MΩ 1/4W resistor |
| R3 | 330Ω 1/4W resistor |
| S1, S2 | SPST switch |

**Table 6-8 Intermediate Signals in Fig. 6-14.**

| **Input A** | | **0** | **0** | **1** | **1** |
|---|---|---|---|---|---|
| **Input B** | | **0** | **1** | **0** | **1** |
| IC1A-3 | ($\overline{B}$) | 1 | 0 | 1 | 0 |
| IC1B-4 | (A NOR $\overline{B}$) | 0 | 1 | 0 | 1 |
| IC1C-10 | ($\overline{A}$) | 1 | 1 | 0 | 0 |
| IC1D-11 | ($\overline{A}$ NOR B) | 0 | 0 | 1 | 0 |
| IC2A-3 | | 1 | 0 | 0 | 1 |
| IC2B-4 | (OUTPUT) | 0 | 1 | 1 | 0 |

majority of inputs are HIGH, the output will be HIGH. By the same token, if the majority of inputs are LOW, the output will be LOW too.

To avoid the problem of tie votes, use an odd number of inputs in a majority-logic circuit. The smallest practical number of inputs to a majority-logic gating circuit is obviously three. For the output to be HIGH, at least two of the inputs must be HIGH. Notice that for a three-input gate, there are eight possible input bit combinations (ranging from 000 to 111). The truth table for a three-input majority-logic gating circuit is as follows:

| INPUTS A B C | OUTPUT |
|---|---|
| 0 0 0 | 0 |
| 0 0 1 | 0 |
| 0 1 0 | 0 |
| 0 1 1 | 1 |
| 1 0 0 | 0 |
| 1 0 1 | 1 |
| 1 1 0 | 1 |
| 1 1 1 | 1 |

This can be reduced effectively to just four generalized possibilities:

All three inputs are LOW
Any two inputs are LOW
Any two inputs are HIGH
All three inputs are HIGH

Notice that all possible bit combinations are covered by this reduction. At least two of the inputs must be at the same state.

You can further reduce the input combinations to just two possibilities. If all three inputs are the same, then at least two must be identical. In other words, the two generalized possibilities are:

Two or more inputs are LOW

or:

Two or more inputs are HIGH

One (and only one) of these two condition statements must be true for all possible input-bit combinations.

A three-input majority-logic gating circuit is shown in Fig. 6-15. A manual input network is shown in Fig. 6-16. An output

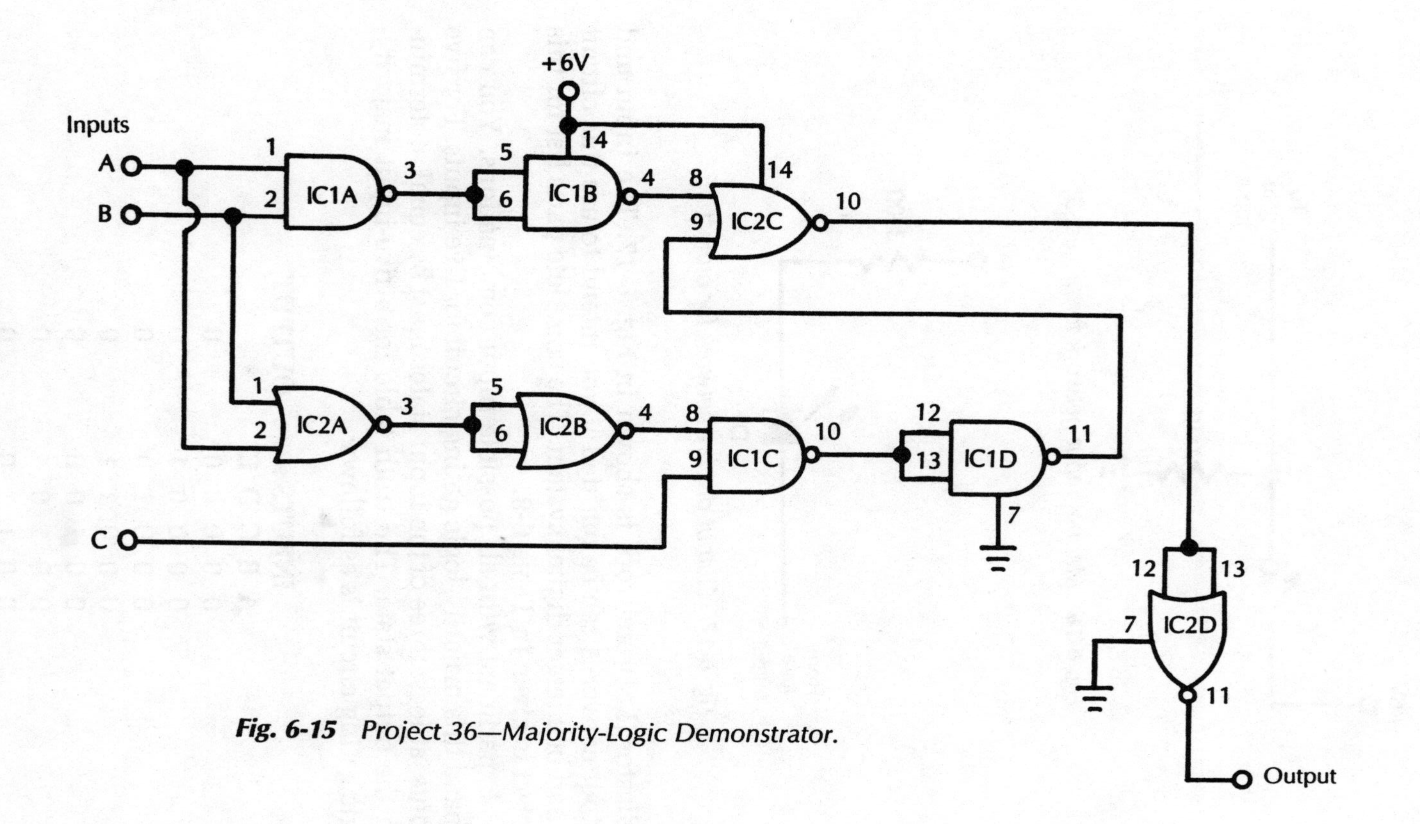


***Fig. 6-15** Project 36—Majority-Logic Demonstrator.*

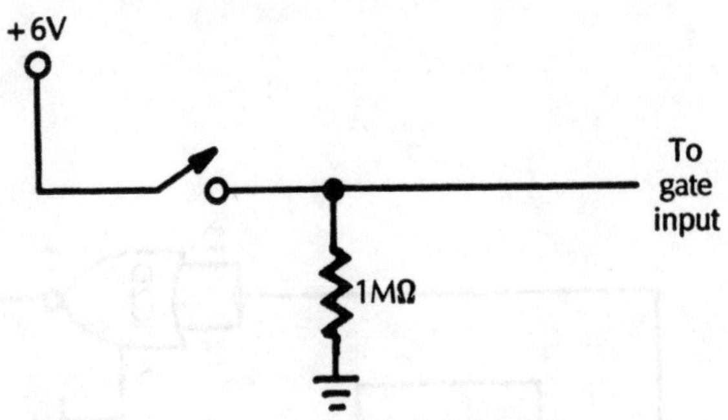


*Fig. 6-16* Manual input network for project 36.

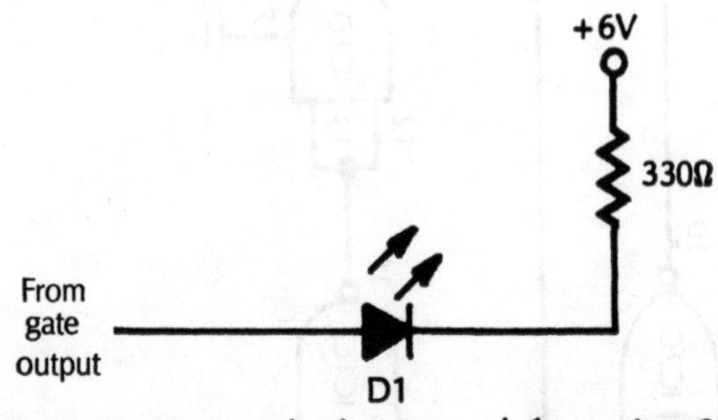


*Fig. 6-17* Output display network for project 36.

indicator (LED) network is shown in Fig. 6-17. The input and output networks can be omitted if you intend to use this circuit as part of a larger digital system. The complete parts list for this project is given in Table 6-9.

Experiment with all possible input combinations. You can expand this majority-logic gating circuit for five inputs. For five inputs, at least three of the input states must be equal, determining the output state. The truth table for a five-input majority-logic gating circuit is as follows:

| INPUTS | | | | | OUTPUT |
|---|---|---|---|---|---|
| A | B | C | D | E | |
| 0 | 0 | 0 | 0 | 0 | 0 |
| 0 | 0 | 0 | 0 | 1 | 0 |
| 0 | 0 | 0 | 1 | 0 | 0 |
| 0 | 0 | 0 | 1 | 1 | 0 |
| 0 | 0 | 1 | 0 | 0 | 0 |
| 0 | 0 | 1 | 0 | 1 | 0 |
| 0 | 0 | 1 | 1 | 0 | 0 |

| | | | | | |
|---|---|---|---|---|---|
| 0 | 0 | 1 | 1 | 1 | 1 |
| 0 | 1 | 0 | 0 | 0 | 0 |
| 0 | 1 | 0 | 0 | 1 | 0 |
| 0 | 1 | 0 | 1 | 0 | 0 |
| 0 | 1 | 0 | 1 | 1 | 1 |
| 0 | 1 | 1 | 0 | 0 | 0 |
| 0 | 1 | 1 | 0 | 1 | 1 |
| 0 | 1 | 1 | 1 | 0 | 1 |
| 0 | 1 | 1 | 1 | 1 | 1 |
| 1 | 0 | 0 | 0 | 0 | 0 |
| 1 | 0 | 0 | 0 | 1 | 0 |
| 1 | 0 | 0 | 1 | 0 | 0 |
| 1 | 0 | 0 | 1 | 1 | 1 |
| 1 | 0 | 1 | 0 | 0 | 0 |
| 1 | 0 | 1 | 0 | 1 | 1 |
| 1 | 0 | 1 | 1 | 0 | 1 |
| 1 | 0 | 1 | 1 | 1 | 1 |
| 1 | 1 | 0 | 0 | 0 | 0 |
| 1 | 1 | 0 | 0 | 1 | 1 |
| 1 | 1 | 0 | 1 | 0 | 1 |
| 1 | 1 | 0 | 1 | 1 | 1 |
| 1 | 1 | 1 | 1 | 1 | 1 |

**Table 6-9 Parts List for Project 36—Majority-Logic Demonstrator.**

| Part | Component Needed |
|---|---|
| | **Demonstrator Fig. 6-15** |
| IC1 | CD4011 quad NAND gate |
| IC2 | CD4001 quad NOR gate |
| | **Manual Input Network (X 3) (Fig. 6-16)** |
| s | SPST switch |
| r | 1MΩ 1/4W resistor |
| | **Output Display Network (Fig. 6-17)** |
| d | LED |
| r | 330Ω 1/4W resistor |

Notice that for all possible combinations of the five inputs, at least three of the inputs are in identical states (either HIGH or LOW). There is always a clear winner in the digital election.

## PROJECT 37: FLIP-FLOP DEMONSTRATOR

A *flip-flop* (bistable multivibrator) is another important element of digital electronics. A flip-flop can be built from standard digital gates, but that construction is not covered in this book.

There are three types of multivibrators:

- Monostable—one stable output state
- Astable—no stable output states
- Bistable—two stable output states

A flip-flop is sometimes called a circuit with a memory. It remembers (and holds) its last output state indefinitely unless the power source is interrupted or the circuit is externally triggered. Each time the flip-flop is triggered, it reverses its output state. The operation of a flip-flop is illustrated in Fig. 6-18. Notice that the output signal is at half the frequency of the input signal. Bistable multivibrators are sometimes called *divide-by-two* counters.

A simple flip-flop demonstration circuit is shown in Fig. 6-19, and the parts list for this project is in Table 6-10. Normally-open, push-button switch S1 is used to trigger the flip-flop manually. The trigger input is normally held LOW (through

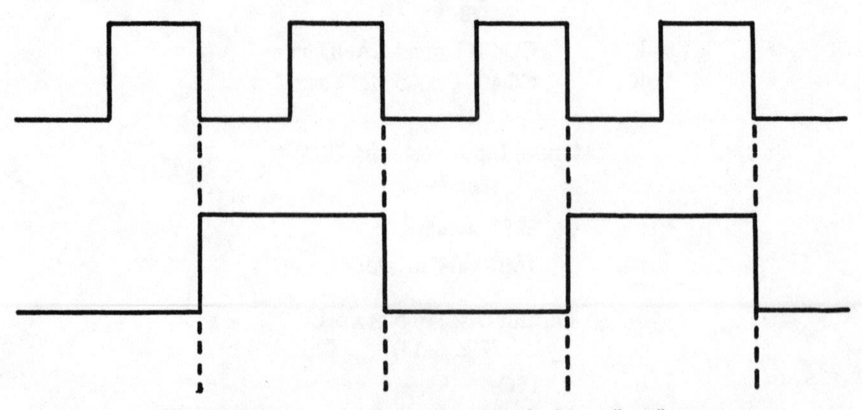

*Fig. 6-18 Input and output signals for a flip-flop.*

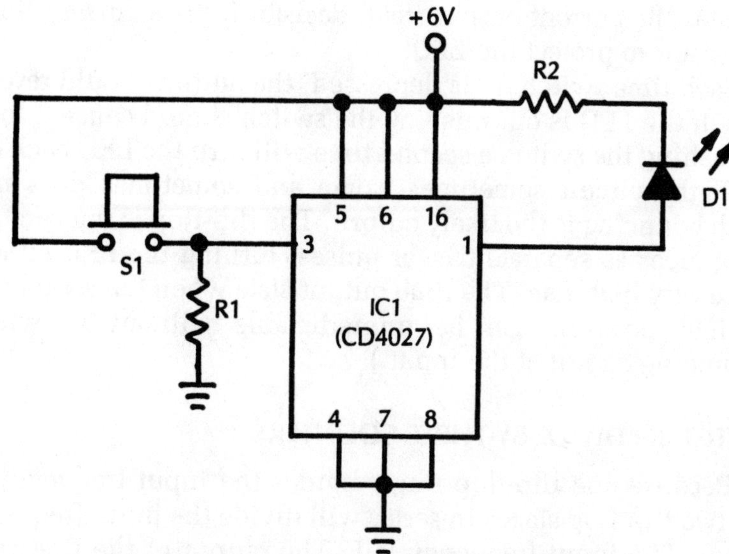


*Fig. 6-19* *Project 37—Flip-Flop Demonstrator.*

**Table 6-10 Parts List for Project 37—Flip-Flop Demonstrator.**

| Part | Component Needed |
|---|---|
| IC1 | CD4027 dual JK flip-flop |
| D1 | LED* |
| R1 | 1MΩ 1/4W resistor* |
| R2 | 330Ω 1/4W resistor* |
| S1 | SPST normally open push-button switch* |

* See text.

resistor R1 to ground) unless the switch is closed. Closing the switch shorts the trigger input to V+, for a HIGH input state.

For reliable results, use a good quality push-button switch for S1. An inexpensive switch might cause problems and incorrect operation due to bouncing. You might want to include a switch-debouncer stage. Refer back to project 28.

If you use an external input signal source to trigger the flip-flop, switch S1 and resistor R1 can be eliminated. The LED (D1) indicates the current output state. Resistor R2 is a current-limiting resistor to protect the LED.

Each time switch S1 is depressed, the output should reverse states. If the LED is on, pushing the switch should cause it to go off. Pushing the switch a second time will turn the LED back on.

(If the circuit sometimes works and sometimes does not, switch bouncing is the likely culprit. The flip-flop is interpreting the bounces as separate trigger pulses, turning the LED on and off at a very high rate. The final output state when the switch settles into position can be unpredictable without a switch-debouncing circuit at the input.)

## PROJECT 38: DIVIDE-BY-THREE COUNTER

Because one flip-flop stage divides the input frequency by two, two flip-flop stages in series will divide the input frequency by four. The input frequency is F. The output of the first stage divides this by two, giving a frequency of F/2. The second stage divides F/2 by two, resulting in a final output frequency of F/4.

That is simple enough. But suppose you need to divide the input signal by a value that is not a power of two. This can be done by forcibly resetting the flip-flop counters before they reach their maximum count.

A divide-by-three counter circuit is shown in Fig. 6-20. Because this circuit calls for nothing other than the CD4027 dual JK flip-flop IC (IC1), there is no need for a parts supply for this project. If you use a switch to trigger the circuit manually, a switch debouncer might be necessary. Refer back to project 28.

## PROJECT 39: DIVIDE-BY-FIVE COUNTER

The circuit shown in Fig. 6-21 carries the principles of project 38 one step further. The output signal has a frequency that is one-fifth of the input frequency. You need two identical ICs for this project. They are CD4027 dual JK flip-flop chips. Because no other components are used, no parts list is given for this project. If you use a switch to trigger the circuit manually, a switch debouncer may be necessary. Refer back to project 28.

The principles used in the last two projects can be expanded for almost any whole number division factor. For large division factors, you will find it more convenient to use a decade counter like the CD4017. You can use a CD4017 in project 40.

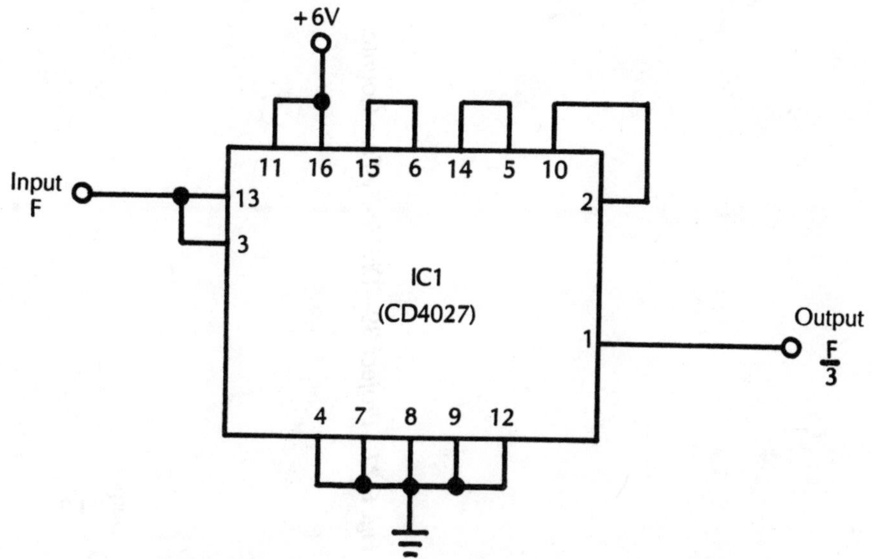


***Fig. 6-20*** *Project 38—Divide-by-Three Counter.*

## PROJECT 40: DIVIDE-BY-130 COUNTER

A string of flip-flop counters is fine for dividing by small integers. But for larger division values, this method rapidly becomes exceedingly unwieldy. Fortunately, dedicated ICs make the task far simpler. (Inside the IC, there is the same type of circuitry you have been dealing with—it has just miniaturized and made more convenient to use.)

A decade counter can divide by any value up to nine. You can add stages in cascade to obtain higher count values. Use the CD4017 decade counter/decoder IC. The pin-out diagram for this chip appears in Fig. 6-22. As the decade counter is triggered by clock pulses, it sequentially provides one of ten outputs. All ten outputs are normally LOW. Only the output indicating the current count value is HIGH. A CARRY OUTPUT (pin #12) is provided to simplify cascading multiple units.

The circuit shown in Fig. 6-23 divides by 130. Three CD4017 decade counters (IC1 – IC3) are used, one for each column in the division value: ones, tens, and hundreds. IC4 is a gate to drive the system output HIGH when the appropriate count outputs are HIGH on all three counter ICs.

Pin #15 on the counter ICs is a reset pin. Ordinarily, resistor R1 holds these pins LOW by providing a current path to ground.

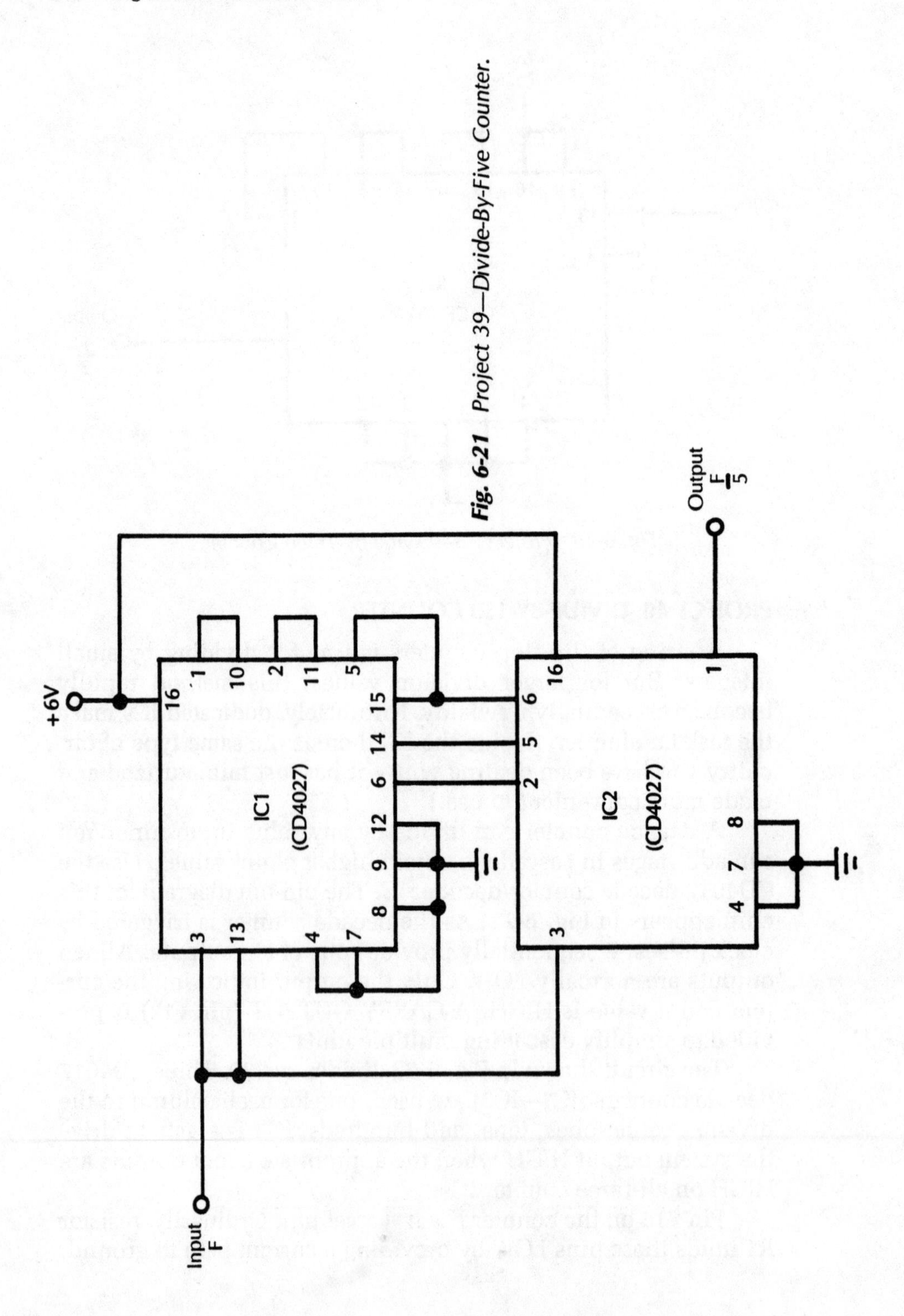


***Fig. 6-21*** *Project 39—Divide-By-Five Counter.*

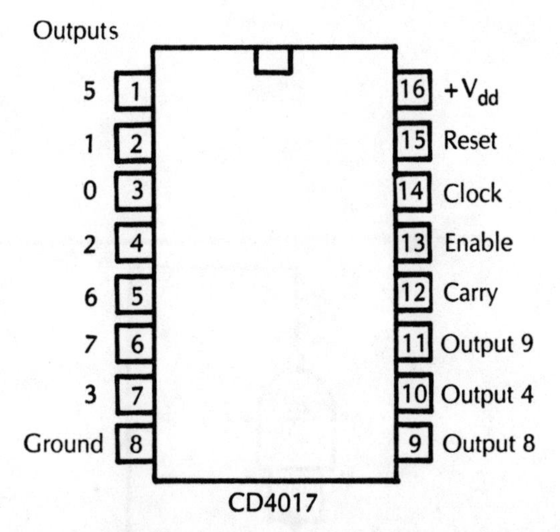


***Fig. 6-22*** *A decade counter is used for large division values.*

When switch S1 (a normally-open, push-button type) is momentarily closed, a HIGH signal is applied to pin #15 of all three counter chips. This resets all counters back to zero.

The parts list for this project is given in Table 6-11. Experiment with other division values, just by feeding different counter outputs to the gates. Any count value up to 999 can be set up with this circuit. For larger count values, add CD4017 counter stages. Connect pin #12 (CARRY OUT) from the last counter IC to pin #14 (CLOCK IN) of the next counter IC in line. Of course, if you use additional counter stages, you must expand the gating network accordingly.

**Table 6-11 Parts List for Project 40—Divide-By-130 Counter.**

| Part | Component Needed |
|---|---|
| IC1, IC2, IC3 | CD4017 decimal counter |
| IC4 | CD4001 quad NAND gate |
| R1 | 1MΩ 1/4W resistor |
| S1 | SPST normally open push-button switch |

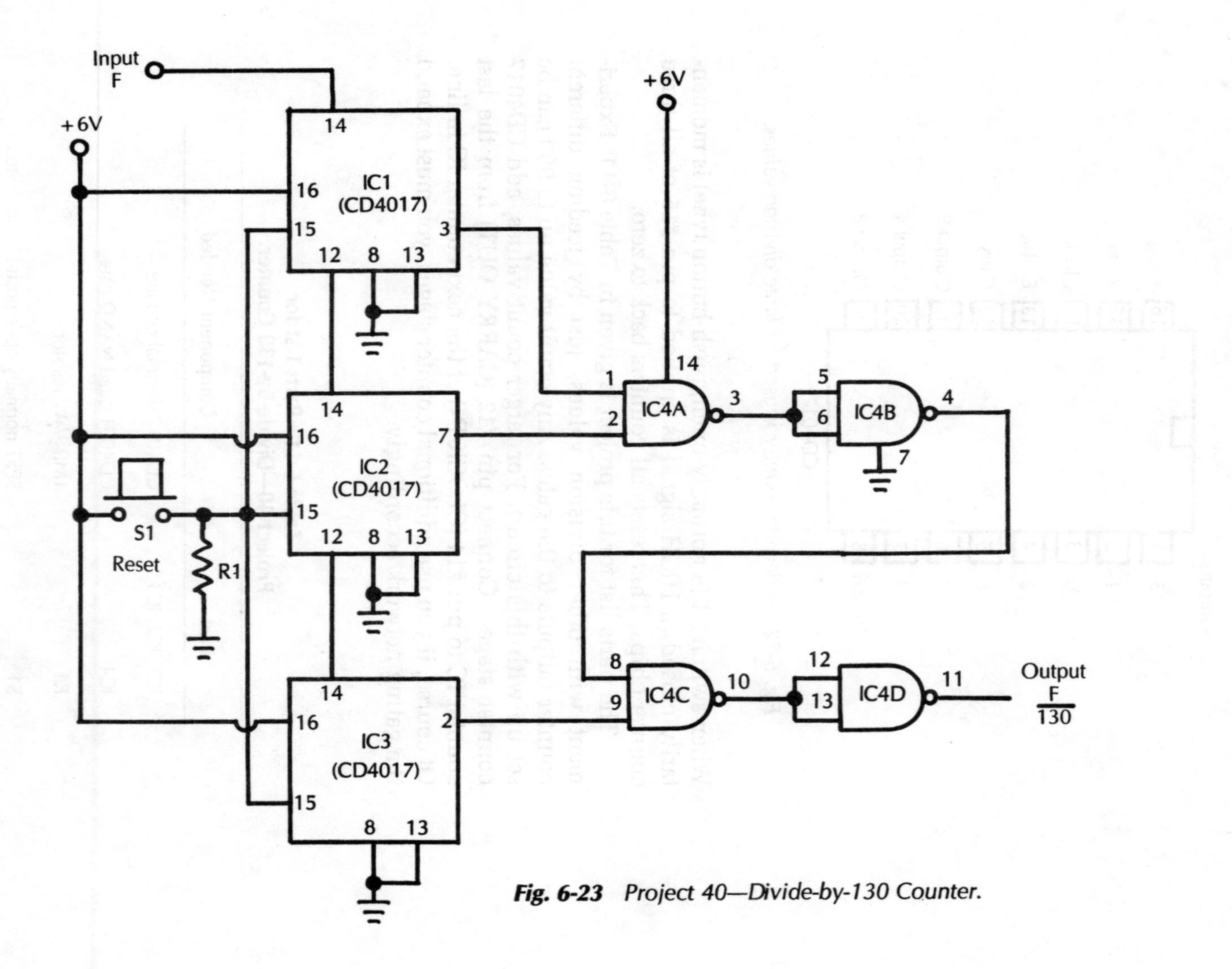


***Fig. 6-23*** *Project 40—Divide-by-130 Counter.*

## PROJECT 41: EIGHT-STEP BINARY COUNTER

You have been using flip-flops to perform frequency division. It is just a small step from this application to a counter. In fact, the term *counter* is used throughout. The major difference between a division circuit and a counter is in whether or not the intermediate outputs are used.

Figure 6-24 shows an eight-step binary counter circuit. Just two ICs are used in this project (CD4013 dual D flip-flops), so there is no point in including a parts list.

Each time a clock pulse is received at the input, the count will be advanced one step. The count sequence works like this:

| OUTPUTS D C B A | DECIMAL COUNT VALUE |
|---|---|
| 0 0 0 0 | 0 |
| 0 0 0 1 | 1 |
| 0 0 1 0 | 2 |
| 0 0 1 1 | 3 |
| 0 1 0 0 | 4 |
| 0 1 0 1 | 5 |
| 0 1 1 0 | 6 |
| 0 1 1 1 | 7 |
| 1 0 0 0 | 8 |
| 1 0 0 1 | 9 |
| 1 0 1 0 | 10 |
| 1 0 1 1 | 11 |
| 1 1 0 0 | 12 |
| 1 1 0 1 | 13 |
| 1 1 1 0 | 14 |
| 1 1 1 1 | 15 |
| 0 0 0 0 | 16 / 0 (the counter resets) |
| 0 0 0 1 | 1 |
| 0 0 1 0 | 2 |
| 0 0 1 1 | 3 |
| 0 1 0 0 | 4 |

and so forth. This sequence will continue repeating as long as power is supplied to the circuit and clock pulses are fed to the input. The triggering clock pulses do not have to be regularly spaced.

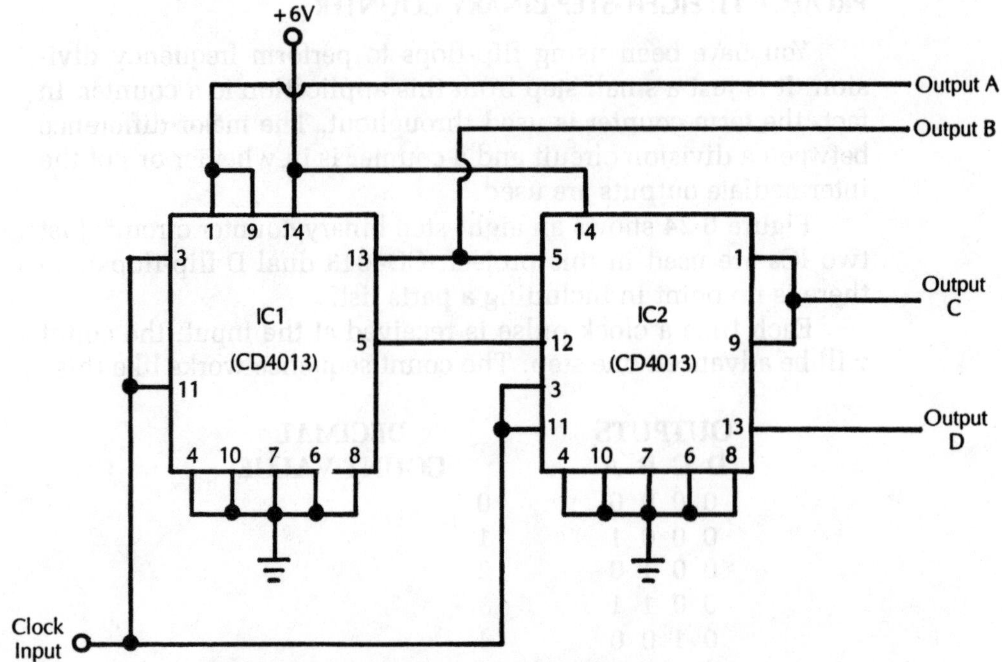


*Fig. 6-24 Project 41—Eight-Step Binary Counter.*

## PROJECT 42: SHIFT REGISTER DEMONSTRATOR

Shift registers are closely related to counters. They are used for short-term memory or to step through a programmed sequence.

A complete, manually controlled shift register project is illustrated in Fig. 6-25, and the parts list for this project is in Table 6-12. Use external input signals in place of the switches if you desire. Similarly, you can use the outputs to drive loads other than just the simple LEDs shown here.

In this circuit, a bit pattern (a string of 1s and 0s) is fed into the data input serially. One bit is accepted by the shift register on each clock pulse. In the manual version, open switch S1 for a LOW input, or close this switch for a HIGH input. You can take as long as you want setting this switch. The circuit will ignore the data switch setting until it is triggered by a clock pulse. Briefly closing the clock switch (S2) will cause the shift register to input the data from switch S1. A switch debouncer might be necessary for switch S2. Refer back to project 28.

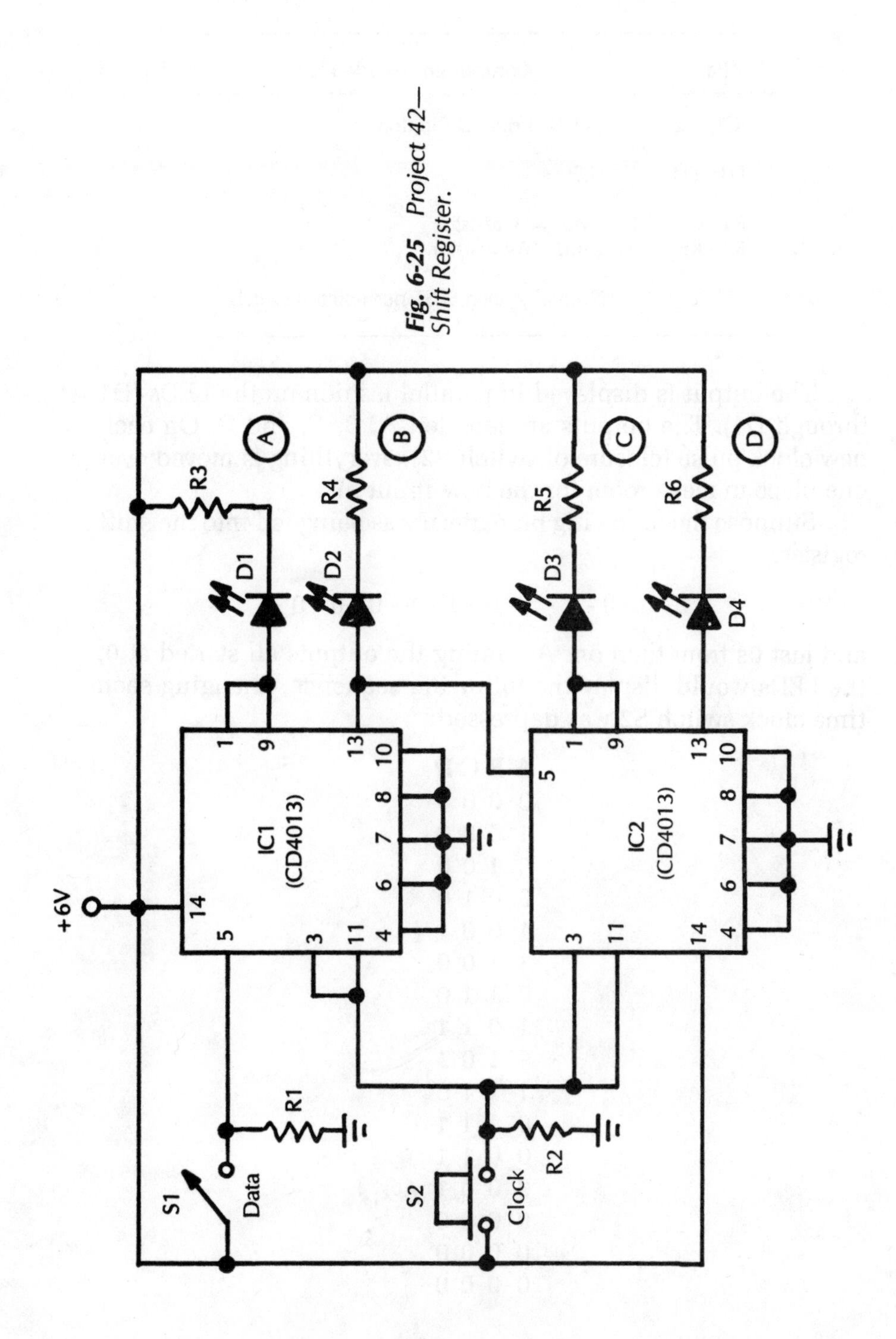


***Fig. 6-25*** *Project 42—Shift Register.*

**Table 6-12 Parts List for Project 42—Shift Register.**

| Part | Component Needed |
|---|---|
| IC1, IC2 | CD4013 dual D flip-flop |
| D1 – D4 | LED |
| R1, R2<br>R3 – R6 | 1MΩ 1/4W resistor<br>330Ω 1/4W resistor |
| S1, S2 | Normally open SPST push-button switch |

The output is displayed in parallel fashion on the LEDs (D1 through D4). The outputs are labelled *A*, *B*, *C*, and *D*. On each new clock pulse (closure of switch S2), everything is moved over one place to make room for the new input bit.

Suppose the following bit pattern was being fed into the shift register:

1 - 0 - 0 - 1 - 1 - 0 - 1 - 1 - 0 - 0 - 0 -

and just 0s from then on. Assuming the outputs all started at 0, the LEDs would display the following sequence, changing each time clock switch S2 was depressed:

| A | B | C | D |
|---|---|---|---|
| 0 | 0 | 0 | 0 |
| 1 | 0 | 0 | 0 |
| 0 | 1 | 0 | 0 |
| 0 | 0 | 1 | 0 |
| 1 | 0 | 0 | 1 |
| 1 | 1 | 0 | 0 |
| 0 | 1 | 1 | 0 |
| 1 | 0 | 1 | 1 |
| 1 | 1 | 0 | 1 |
| 1 | 1 | 1 | 0 |
| 0 | 1 | 1 | 1 |
| 0 | 0 | 1 | 1 |
| 0 | 0 | 0 | 1 |
| 0 | 0 | 0 | 0 |
| 0 | 0 | 0 | 0 |
| 0 | 0 | 0 | 0 |

Because only 0s are now being entered into the shift register, there will be no further changes in the outputs. The shifted data might also be taken off serially at output D, delayed by the number of stages in the register. In this project, there are four stages. Output D will be four clock pulses behind the input.

## PROJECT 43: SIMPLE LOGIC PROBE

If you work much with digital circuits, you will certainly need a logic probe. This is a simple piece of digital test equipment. It is connected to the V+ and ground connection points of the circuit under test, thereby stealing its power source. You can touch a probe to any point within a digital circuit. The logic probe will tell you what the logic state is at that point.

A simple, but effective logic-probe circuit appears in Fig. 6-26, and the parts list for this project is given in Table 6-13. Nothing is terribly critical in this circuit. The NAND gates are wired as inverters. Substitute NOR gates, or dedicated inverters, if you wish. The values of the resistors control the brightness of the LEDs.

Two LEDs are used to display the logic state at the probe tip. If LED D1 is lit and LED D2 is dark, then the logic state is HIGH. If LED D2 is lit and LED D1 is dark, then the logic state is LOW. If both LEDs appear to be lit simultaneously (perhaps at a slightly lower than normal intensity), then a high frequency pulse stream is being detected. The LEDs are actually alternately flashing on and off too rapidly for your eyes to catch the individual blinks.

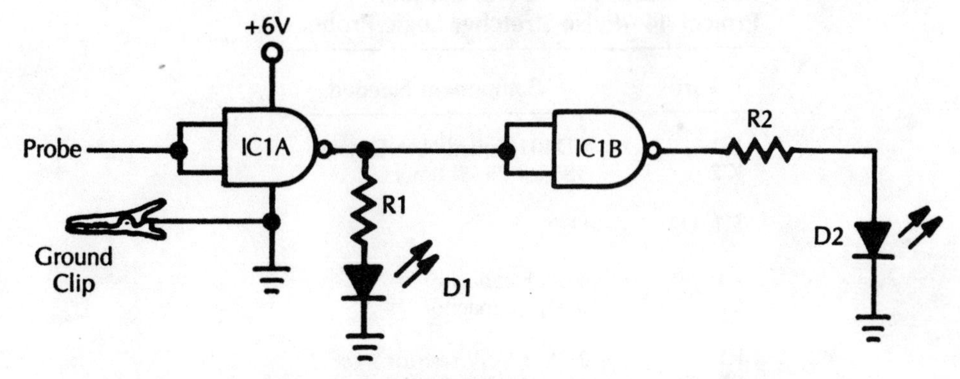


***Fig. 6-26*** *Project 43—Logic Probe.*

**Table 6-13 Parts List for Project 43—Logic Probe.**

| Part | Component Needed |
|---|---|
| IC1 | CD4011 quad NAND gate |
| D1, D2 | LED |
| R1, R2 | 470Ω 1/4W resistor |

## PROJECT 44: PULSE-STRETCHER LOGIC PROBE

Sometimes when working with a digital circuit, you need to watch for a brief, isolated pulse. It might happen too fast for you to see the LED flash. You might blink just as the LED flashes.

The solution is to stretch out the pulse so that you can easily see the glow of the LED. A circuit for accomplishing this is shown in Fig. 6-27, and the parts list for this project is given in Table 6-14.

Basically, this is the simple logic-probe circuit of project 43 with a built-in 555 timer monostable multivibrator. The timer serves as a pulse stretcher. For more information on how this part of the circuit works, refer back to project 22.

You might want to experiment with alternate values for the timing components—resistor R1, and capacitor C1.

**Table 6-14 Parts List for Project 44—Pulse-Stretcher Logic Probe.**

| Part | Component Needed |
|---|---|
| IC1 | CD4011 quad NAND gate |
| IC2 | 555 (or 7555) timer |
| D1, D2 | LED |
| C1 | 0.68μF capacitor |
| C2 | 0.01μF capacitor |
| R1 | 2.2MΩ 1/4W resistor |
| R2, R3 | 470Ω 1/4W resistor |

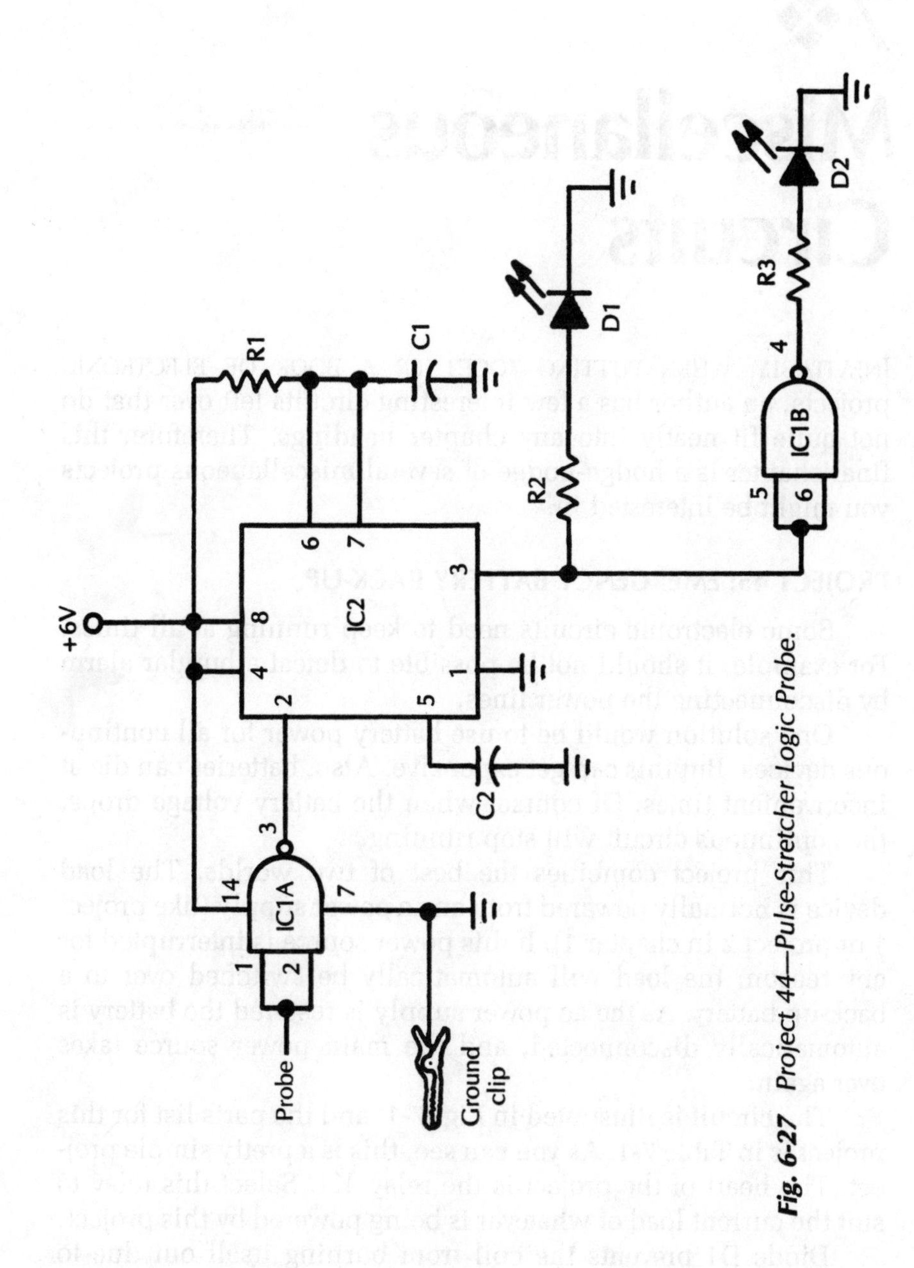


**Fig. 6-27** *Project 44—Pulse-Stretcher Logic Probe.*

# 7 ❖ Miscellaneous Circuits

INEVITABLY, WHEN PUTTING TOGETHER A BOOK OF ELECTRONIC projects, an author has a few interesting circuits left over that do not quite fit neatly into any chapter headings. Therefore, this final chapter is a hodge-podge of several miscellaneous projects you might be interested in.

## PROJECT 45: EMERGENCY BATTERY BACK-UP

Some electronic circuits need to keep running at all times. For example, it should not be possible to defeat a burglar alarm by disconnecting the power lines.

One solution would be to use battery power for all continuous devices. But this can get expensive. Also, batteries can die at inconvenient times. Of course, when the battery voltage drops, the continuous circuit will stop running.

This project combines the best of two worlds. The load device is normally powered from an ac power supply (like project 1 or project 2 in chapter 1). If this power source is interrupted for any reason, the load will automatically be switched over to a back-up battery. As the ac power supply is restored the battery is automatically disconnected, and the main power source takes over again.

The circuit is illustrated in Fig. 7-1, and the parts list for this project is in Table 7-1. As you can see, this is a pretty simple project. The heart of the project is the relay, K1. Select this relay to suit the current load of whatever is being powered by this project.

Diode D1 prevents the coil from burning itself out due to back EMF. Use almost any silicon diode here.

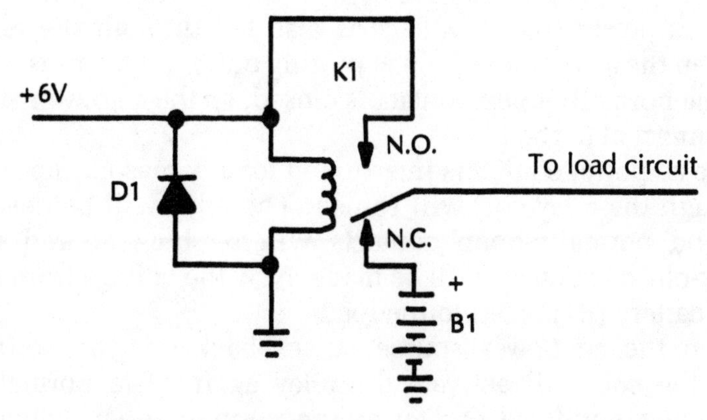


***Fig. 7-1*** *Project 45—Emergency Battery Back-Up.*

**Table 7-1 Parts List for Project 45—Emergency Battery Back-Up.**

| Part | Component Needed |
|---|---|
| D1 | 1N4001 diode (or any silicon diode) |
| K1 | 6V relay with SPDT contacts. Current rating must suit the load. |
| B1 | 6V battery (back-up) |

The relay contacts should be of the *SPDT* (single-pole, double-throw) type. There are three contacts: a center-common contact, a normally-open contact, and a normally-closed contact. When the relay is deactivated, the normally-open contact is open and the normally-closed contact is closed. But when the relay is activated by a sufficient voltage flowing through its coil, the situation is reversed. The normally-open contact is now closed, and the normally-closed contact is opened.

Make the following connections to the relay switching contacts:

| RELAY CONTACT | CONNECTION |
|---|---|
| Common (center) | Load |
| Normally Open | ac Power Source |
| Normally Closed | Back-Up Battery (B1) |

The ac power source voltage is also fed through the relay coil. When the ac power supply is putting out 6V, the coil is activated. The normally-open contact is closed, so the ac power supply is connected to the load.

If the ac power source is interrupted for any reason, the voltage through the relay coil will vanish. The relay will be deactivated. The normally-open contacts will be broken, and the normally-closed contacts will be made. Now the voltage from the back-up battery (B1) is fed to the load.

When the ac power source comes back on, the voltage through the coil will activate the relay again. The normally-closed contacts will be broken as the normally-open contacts close. The load device will continue operating even though the ac power source is interrupted.

This project is very simple, and might not work well in some applications, especially those involving computers and certain other digital circuits. The relay contacts do not switch instantly—just very rapidly. There might be enough of a delay to cause a serious glitch in digital data, or even to cause *RAMs* (random-access memories) to clear. But for less-critical applications, this project will do the job very nicely.

Ordinarily, the battery (B1) is just sitting there, so there is not drain on it. If the ac power source is never interrupted, this battery will have its full shelf life. Of course, if the battery is connected to the load, its lifespan will depend on the amount of time the battery must power the load, and the current drain of the load.

## PROJECT 46: ELECTRONIC COIN FLIPPER

This project is a just-for-fun project you will enjoy. It performs the equivalent of electronically flipping a coin. This project might also be called an executive decision maker. The schematic diagram for this project appears in Fig. 7-2, and the parts list is in Table 7-2.

IC1 (a 555 or 7555 timer) is wired as a high-speed oscillator. When push-button switch S1 is held closed, the oscillator output pulses are fed into a flip-flop (IC2), which functions as a latch.

When switch S1 is released (opened), the latch (IC2) remembers and holds its last state (HIGH or LOW). If the latch output is HIGH, LED D1 will be lit; otherwise it will be dark.

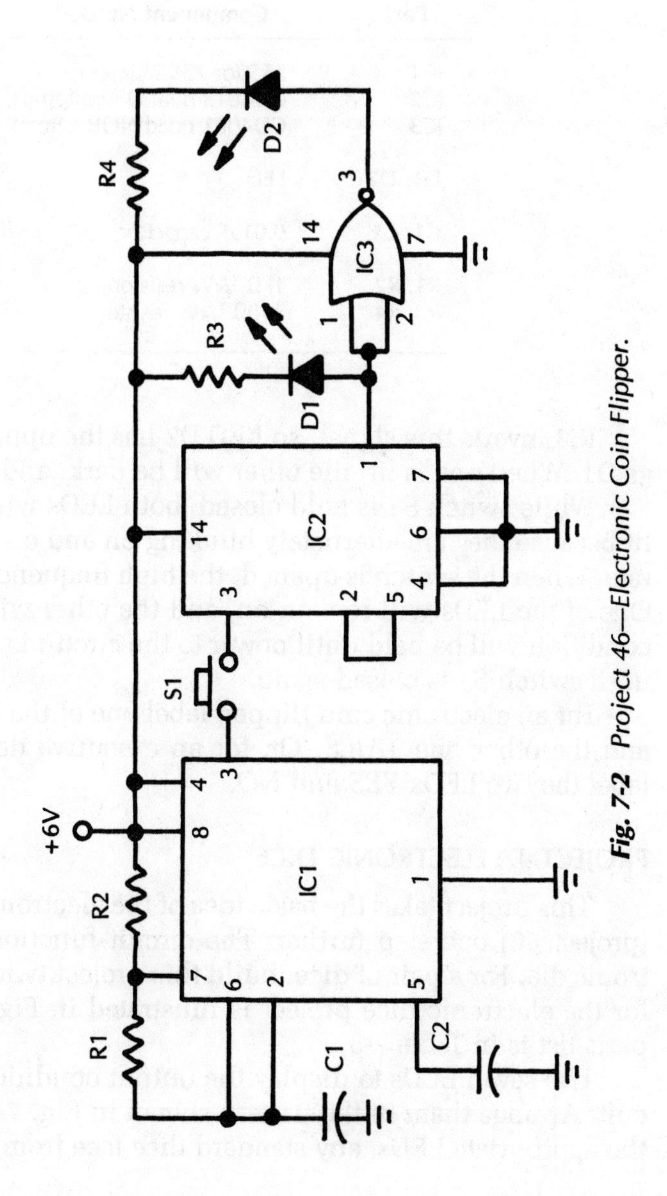


***Fig. 7-2*** *Project 46—Electronic Coin Flipper.*

**Table 7-2 Parts List for Project 46—Electronic Coin Flipper.**

| Part | Component Needed |
|---|---|
| IC1 | 555 (or 7555) timer |
| IC2 | CD4013 dual D flip-flop |
| IC3 | CD4001 quad NOR gate |
| D1, D2 | LED |
| C1, C2 | 0.01$\mu$F capacitor |
| R1, R2 | 1k$\Omega$ $^1/_4$W resistor |
| R3, R4 | 330$\Omega$ $^1/_4$W resistor |

IC3 inverts this signal, so LED D2 has the opposite response as D1. When one is lit, the other will be dark, and vice versa.

While switch S1 is held closed, both LEDs will appear to be lit because they are alternately blinking on and off at a very high rate. When the switch is opened, the high frequency pulses stop. One of the LEDs will remain on, and the other will go off. This condition will be held until power to the circuit is interrupted or until switch S1 is closed again.

For an electronic coin flipper, label one of the LEDs *HEADS*, and the other one *TAILS*. Or, for an executive decision maker, label the two LEDs *YES* and *NO*.

## PROJECT 47: ELECTRONIC DICE

This project takes the basic idea of the electronic coin flipper (project 46) one step further. The circuit functions as an electronic die. For a pair of dice, build this project twice. The circuit for the electronic dice project is illustrated in Fig. 7-3, and the parts list is in Table 7-3.

Use seven LEDs to display the output conditions of this circuit. Arrange them in the pattern shown in Fig. 7-4. By lighting the appropriate LEDs, any standard dice face from one to six can

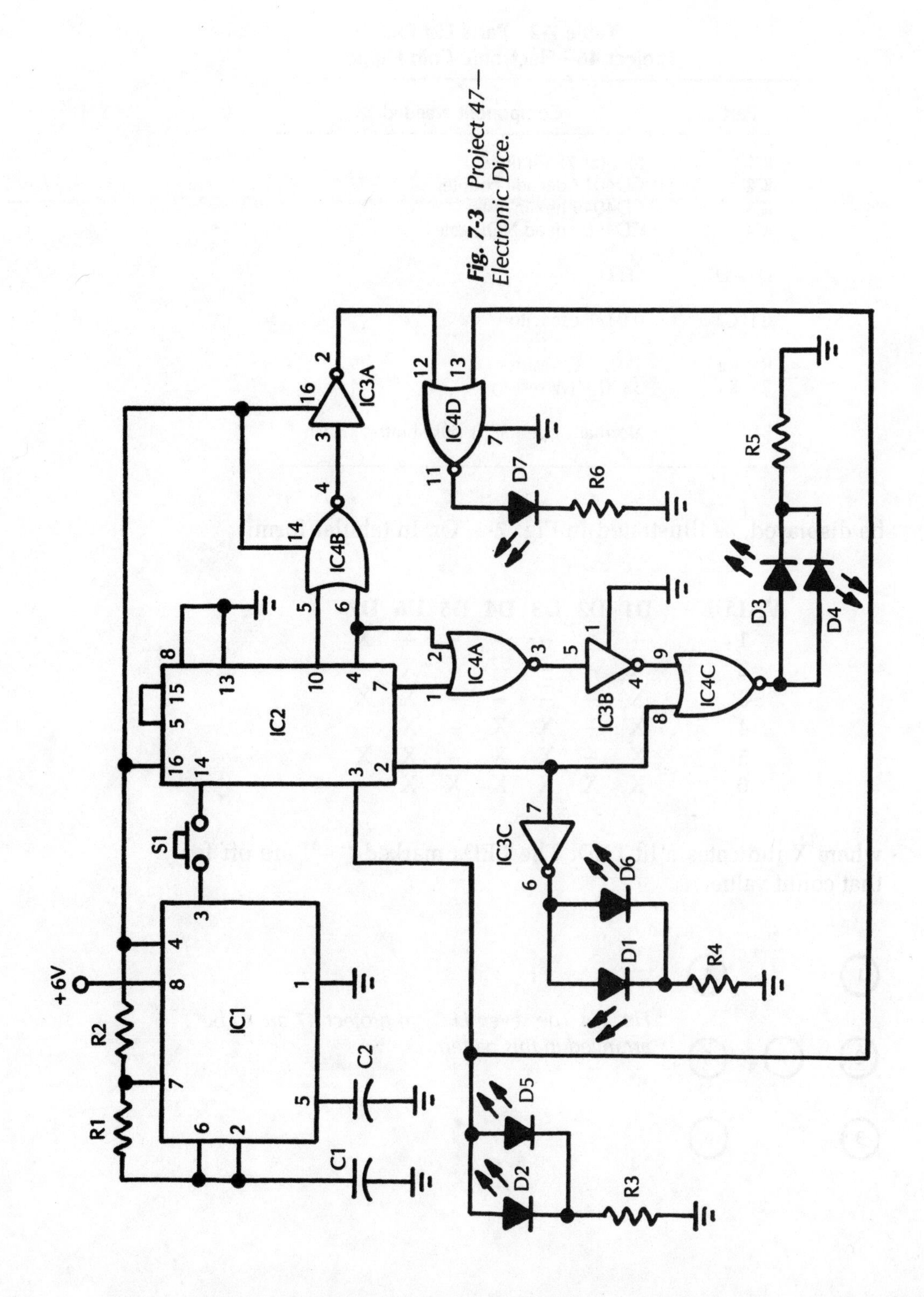


***Fig. 7-3** Project 47—Electronic Dice.*

**Table 7-2 Parts List for Project 46—Electronic Coin Flipper.**

| Part | Component Needed |
|---|---|
| IC1 | 555 (or 7555) timer |
| IC2 | CD4017 decade counter |
| IC3 | CD4049 hex inverter |
| IC4 | CD4001 quad NOR gate |
| D1 – D7 | LED |
| C1, C2 | 0.01$\mu$F capacitor |
| R1, R2 | 1kΩ 1/4W resistor |
| R3, R4 | 330Ω 1/4W resistor |
| S1 | Normally open SPST push-button switch |

be displayed, as illustrated in Fig. 7-5. Or, in tabular form:

| VALUE | D1 | D2 | D3 | D4 | D5 | D6 | D7 |
|---|---|---|---|---|---|---|---|
| 1 | – | – | – | – | – | – | X |
| 2 | X | – | – | – | – | X | – |
| 3 | X | – | – | – | – | X | X |
| 4 | X | – | X | X | – | X | – |
| 5 | X | – | X | X | – | X | X |
| 6 | X | X | X | X | X | X | – |

where $X$ indicates a lit LED. The LEDs marked "–" are off for that count value.

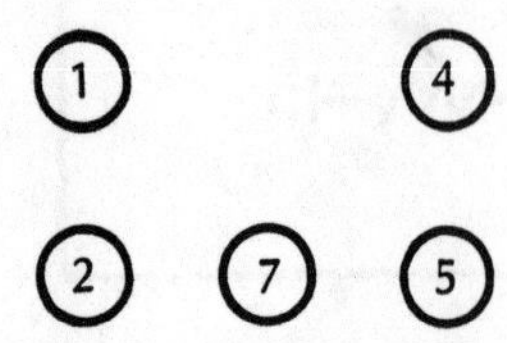

*Fig. 7-4 The seven LEDs in project 47 are to be arranged in this pattern.*

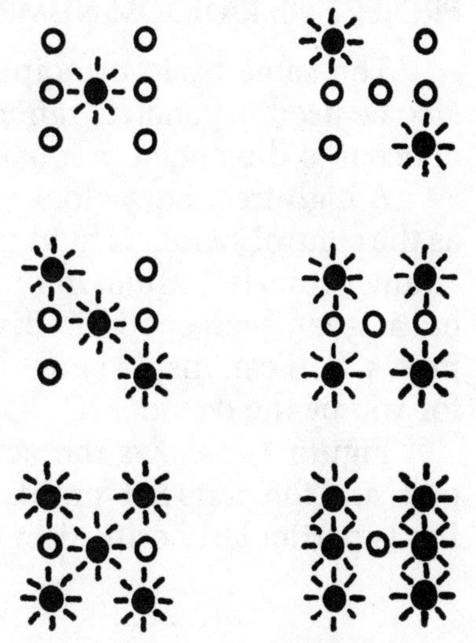

*Fig. 7-5* Any standard dice-face pattern can be displayed on project 47.

Notice that certain LED pairs are always used together. Either they are both dark, or they are both lit;

| | |
|---|---|
| D1 | D6 |
| D2 | D5 |
| D3 | D4 |

These diodes might be wired in parallel, simplifying the gating network in the circuit.

IC1 (555 or 7555 timer) is a high-frequency clock. When switch S1 is open (its normal condition), no pulses can get through to the next stage. Closing switch S1 permits the pulses to advance a decade counter (IC2). This counter is wired for a six-step count. IC3 and IC4 gate the count outputs to light up the appropriate LEDs for each of the six possible count values.

None of the component values are critical in this circuit. If any of the frequency determining components (capacitors C1, resistors R1 and R2) are given values that are too large, you will be able to see the various patterns flash across the display as the counter is being incremented. With a suitably high frequency, all seven LEDs will appear to be simultaneously lit, making it impossible to predict what final value will be displayed when switch S1 is released.

## PROJECT 48: RANDOM-NUMBER GENERATOR

The same basic principles used in projects 46 and 47 can also be used to generate random decimal values from 0 to 9. The concept is illustrated in block diagram form in Fig. 7-6.

A high-frequency clock rapidly increments a counter as long as the control switch is held closed. Opening the switch stops the counter at its last value. In this case, the output value is indicated on a seven-segment LED display. By lighting appropriate segments, you can display any digit from 0 to 9. The gating is done for you by the decoder IC (IC3).

Figure 7-7 shows the actual random number generator circuit, and the parts list for this project is in Table 7-4. Again, nothing is particularly critical in this circuit.

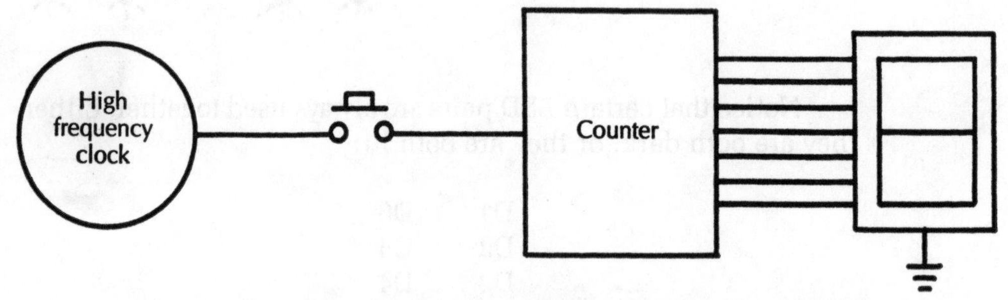


*Fig. 7-6 Project 48—Block Diagram for Random-Number Generator.*

**Table 7-4 Parts List for Project 48—Random-Number Generator.**

| Part | Component Needed |
| --- | --- |
| IC1 | 555 (or 7555) timer |
| IC2 | CD4518 BCD counter |
| IC3 | CD4511 BCD to 7-segment decoder |
| DIS1 | Seven-segment LED display with common cathode |
| C1, C2 | 0.01µF capacitor |
| R1, R2 | 1kΩ ¼W resistor |
| R3–R9 | 330Ω ¼W resistor |
| S1 | Normally open SPST push-button switch |

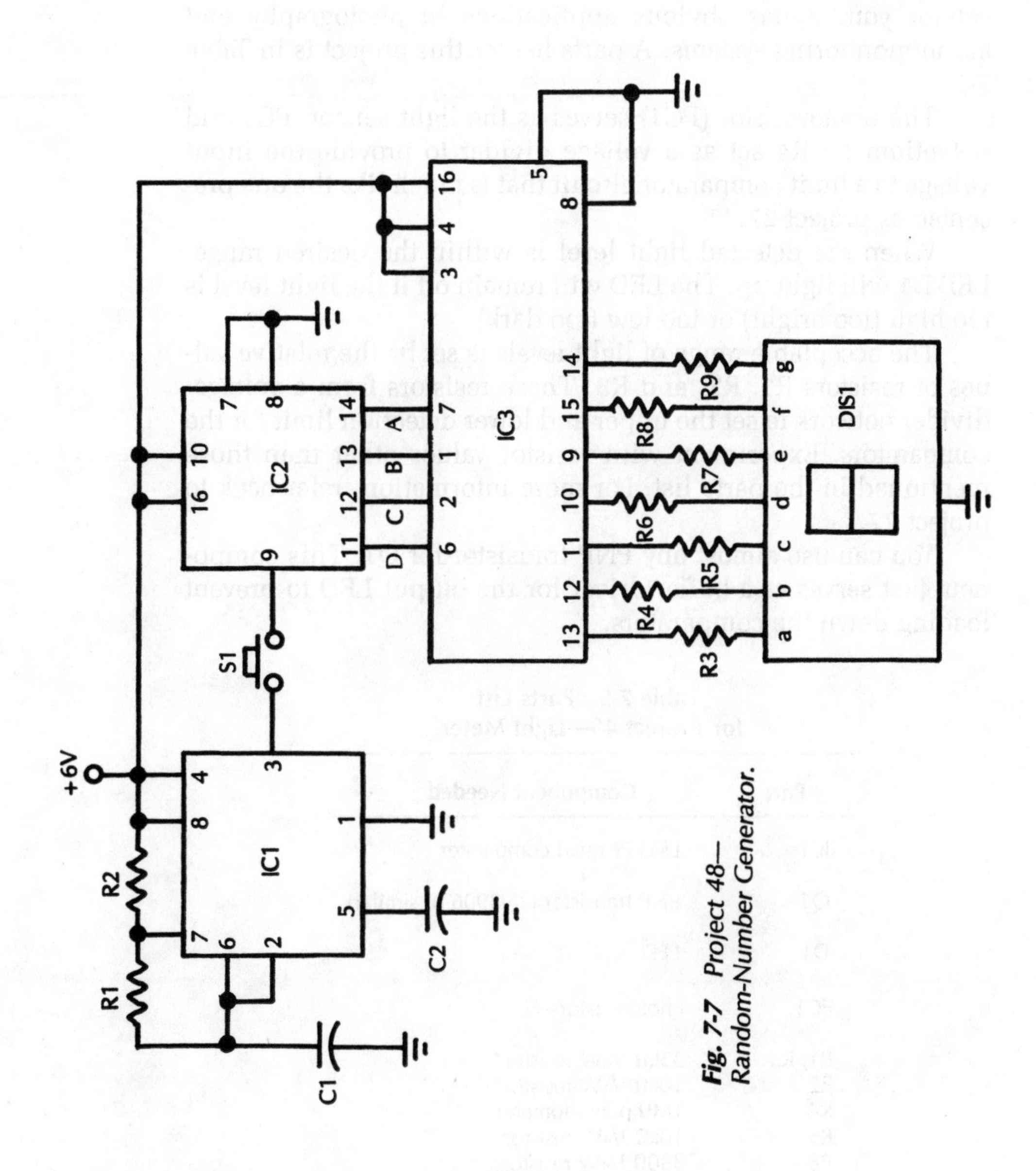


***Fig. 7-7*** *Project 48—*
*Random-Number Generator.*

## PROJECT 49: LIGHT METER

Need to know whether the light level is within a certain range? If so, the light meter circuit shown in Fig. 7-8 is the project for you. It has obvious applications in photography and alarm/monitoring systems. A parts list for this project is in Table 7-5.

The photoresistor (PC1) serves as the light sensor. PC1 and potentiometer R4 act as a voltage divider to provide the input voltage to a limit comparator circuit that is much like the one presented as project 27.

When the detected light level is within the desired range, LED D1 will light up. The LED will remain off if the light level is too high (too bright) or too low (too dark).

The acceptable range of light levels is set by the relative values of resistors R1, R2, and R3. These resistors form a voltage-divider network to set the upper and lower detection limit for the comparators. Experiment with resistor values other than those mentioned in the parts list. For more information, refer back to project 27.

You can use almost any PNP transistor for Q1. This component just serves as a buffer/driver for the output LED to prevent loading down the comparators.

**Table 7-5 Parts List for Project 49—Light Meter.**

| Part | Component Needed |
|---|---|
| IC1 | LM339 quad comparator |
| Q1 | PNP transistor (2N3906 or similar) |
| D1 | LED |
| PC1 | Photoresistor |
| R1, R3 | 33kΩ 1/4W resistor* |
| R2 | 10kΩ 1/4W resistor* |
| R4 | 1MΩ potentiometer |
| R5 | 10kΩ 1/4W resistor |
| R6 | 330Ω 1/4W resistor |

*See text.

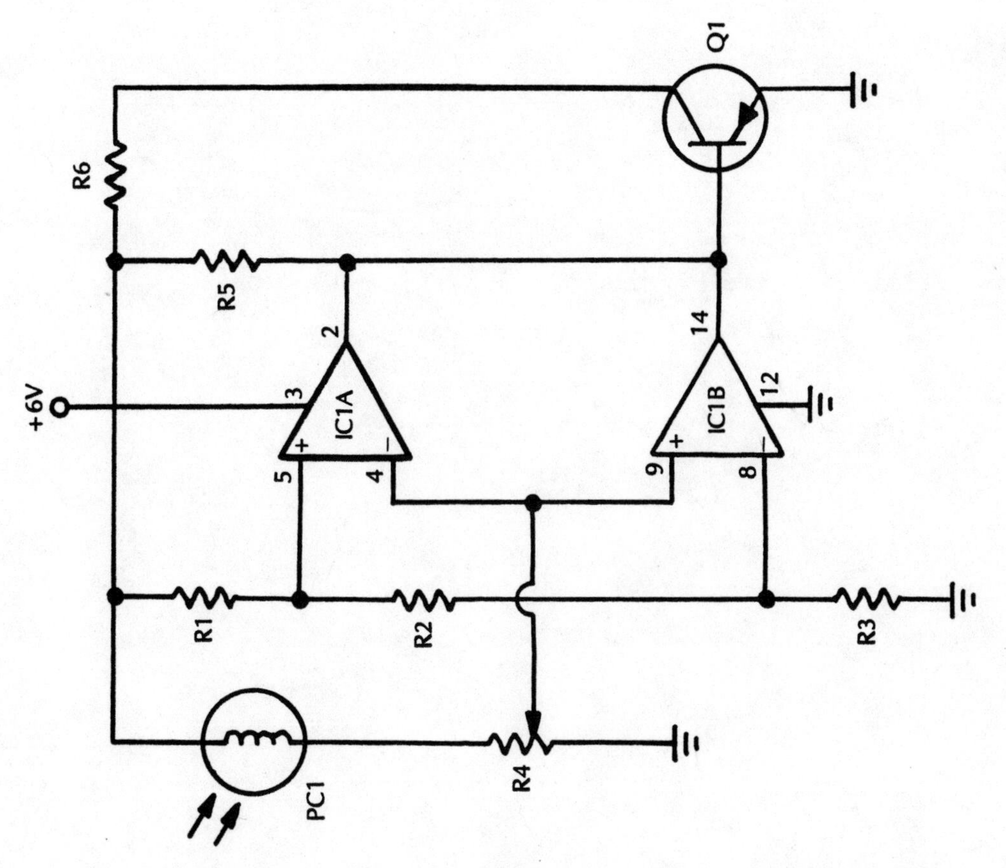


*Fig. 7-8* Project 49—Light Meter.

Adjust potentiometer R4 to set the sensitivity of the photosensor. This control interacts with the values of resistors R1, R2, and R3.

*Well, there you have it; forty-nine simple projects that can be powered by + 6V. I hope you enjoy building and experimenting with these projects as much as I've enjoyed presenting them to you.*

# Index

## A



## B



## C

# R



# S



# T



# U



# V



# W